SOLVING PROBLEMS IN DYNAMICS, ELECTRICITY AND MAGNETISM

E. W. LAING
Senior Lecturer, Department of Natural Philosophy
University of Glasgow

W. McFARLANE
Senior Lecturer, Department of Natural Philosophy
University of Glasgow

OLIVER & BOYD · EDINBURGH

Oliver & Boyd
Tweeddale Court
14 High Street
Edinburgh EH1 1YL

A Division of Longman Group Limited

First published 1972

Set in Times Roman and Printed in
Great Britain by Bell & Bain Ltd.
Glasgow

SOLVING PROBLEMS IN PHYSICS

Series Editor F. W. Laing

3

DYNAMICS, ELECTRICITY AND MAGNETISM

Titles in this series

1 *Mechanics*—R. B. Gardiner and E. W. Laing

2 *Waves and Vibrations*—N. MacDonald

3 *Dynamics, Electricity and Magnetism*—E. W. Laing and W. McFarlane

PREFACE

This book is not intended to be a fully comprehensive collection of solved problems in dynamics, electricity and magnetism. Our primary aim is to help students to solve problems *by themselves*, thereby deepening their understanding of some of the basic principles of physics. It is therefore quite possible that some important topics have been omitted, but we believe that a thorough understanding of those topics chosen for inclusion in this book will lead to a better appreciation of the fundamentals and thereby provide a firmer foundation for further study.

The material is appropriate for Honours classes in physics and has been arranged in two parts, the first dealing with dynamics and the second with electricity and magnetism. Each part is divided into short chapters, with unsolved problems included at the end of each chapter.

We would like to thank the University Court of the University of Glasgow for permission to make use of questions set in recent class and degree examinations.

CONTENTS

CONTENTS

PART 1 DYNAMICS

1 GENERALISED COORDINATES

Generalised coordinates are introduced to take advantage of constraints on a dynamical system which have the effect of reducing the number of degrees of freedom possessed by the system. A generalised coordinate is usually introduced as a measure of the displacement of part of the system from some initial position, and could equally well represent a linear or curvilinear displacement. All such coordinates are to be regarded on an equal footing.

Many students find difficulty in counting up the number of degrees of freedom of a system except in the simplest cases, yet it is a straightforward task. If a system consists of several connected parts, then it is a question of counting the number of parameters required to determine the position of one part. With that done, it is necessary to find how many more parameters are required to fix any part connected to the first, and so on.

Let us consider, for example, a system composed of an arbitrary rigid body connected to another rigid body by means of a rigid rod, both joints being universal joints. The whole system is free to move in space and we shall first discuss the degrees of freedom of one rigid body. A point A in the body requires three parameters, say three cartesian coordinates relative to a fixed set of axes. Once this is done, any other point B in the body is constrained to be at a fixed distance, say r, from the point A but otherwise free. It therefore lies on the surface of a sphere, centre A and radius r. Two parameters, for example a polar and an azimuthal angle, will then determine B. Given A and B, all points of the body on the line AB are fixed, but the body may rotate about AB as axis of rotation. Any other point C in the body is fixed by the angle of rotation of the body relative to a plane through AB fixed in space. From such arguments we may deduce that a rigid body possesses six degrees of freedom.

Next we have to consider the connecting rod. One end is already fixed. The other end clearly requires two more parameters (like the point B of the rigid body). Finally, the second rigid body can be treated like the first, but already has a point on its surface fixed by the rod. Three more parameters are still required. Altogether, the system has eleven degrees of freedom.

We shall in this chaper consider only holonomic dynamical systems, which have the property that the number of *independent* generalised coordinates required to define an instantaneous state of the system is just equal to the number of degrees of freedom of the system. (A useful discussion about non-holonomic systems is given by K. R. Symon in *Mechanics*, published in 1960 by Addison–Wesley).

In all situations, an expression is obtained for the kinetic energy T of the system as a function of the independent generalised coordinates q_j and velocities $\dot{q}_j$. For a system of s degrees of freedom

$$T = T(q_1 \ldots q_s, \dot{q}_1 \ldots \dot{q}_s, t).$$

In many problems, the external forces are derivable from a potential function V,

$$V = V(q_1 \ldots q_s),$$

and Lagrange's equations of motion are

$$\frac{d}{dt}\left(\frac{\partial T}{\partial \dot{q}_j}\right) - \frac{\partial T}{\partial q_j} = -\frac{\partial V}{\partial q_j}, \quad j = 1 \ldots s.$$

Sometimes there is no potential function and the equations of motion have instead a term Q_j on the right hand side. This is the generalised force, best defined in terms of the virtual work δW done by external forces when virtual displacements $\delta q_1 \ldots \delta q_s$ occur consistent with the constraints (note that these virtual displacements are independent of each other). Thus,

$$\delta W = \sum_{j=1}^{s} Q_j \, \delta q_j.$$

$Q_j \, \delta q_j$ always has the dimensions of energy, but unless q_j is a length, Q_j will not have the dimensions of force. What is Q_j if q_j is an angle? Lagrange's equations then take the form

$$\frac{d}{dt}\left(\frac{\partial T}{\partial \dot{q}_j}\right) - \frac{\partial T}{\partial q_j} = Q_j, \quad j = 1 \ldots s.$$

As a simple illustration, let us obtain the equations of motion of a particle moving in a plane under the action of an external force F, in terms of plane polar coordinates r, θ as generalised coordinates.

Let $F = (F_r, F_\theta)$. Now consider a displacement δr with θ fixed. The work done is $F_r\,\delta r$. Next consider $\delta\theta$, with r fixed. In this case the work done is $F_\theta\, r\delta\theta$. Thus for a general displacement $(\delta r, \delta\theta)$

$$\delta W = F_r\,\delta r + rF_\theta\,\delta\theta = Q_r\,\delta r + Q_\theta\,\delta\theta.$$

Hence $F_r = Q_r$ and $rF_\theta = Q_\theta$

$$T = \tfrac{1}{2}m(\dot{r}^2 + r^2\dot{\theta}^2).$$

$$\frac{d}{dt}\left(\frac{\partial T}{\partial \dot{r}}\right) - \frac{\partial T}{\partial r} = Q_r. \quad \text{Hence } m(\ddot{r} - r\dot{\theta}^2) = F_r$$

$$\frac{d}{dt}\left(\frac{\partial T}{\partial \dot{\theta}}\right) - \frac{\partial T}{\partial \theta} = Q_\theta. \quad \text{Hence } \frac{d}{dt}(mr^2\dot{\theta}) = rF_\theta$$

1.1. *A solid uniform circular cylinder of radius a, mass M, lies on a horizontal table with its axis horizontal, and a rod AB of length l, mass m, rests across the cylinder at right angles to its axis, with the end A resting on the table. No slipping occurs between the cylinder and the table, but the rod is perfectly smooth. The system is held stationary and then released. Show that in the subsequent motion the kinetic energy T is given by*

$$T = \tfrac{1}{2}m(\dot{x}^2 + \dot{x}\dot{\theta}l\sin\theta) + \tfrac{1}{6}ml^2\dot{\theta}^2 + \tfrac{3}{4}M\dot{y}^2$$

where θ is the angle between the horizontal and the rod, and x and y are the distances of A and P respectively from the initial position of P, the point of the vertical plane through AB at which the cylinder is in contact with the table.

Hence show that when the end B of the rod first touches the cylinder

$$(2m + 3M)\dot{y} + \frac{m}{a}\left(\frac{a^4 + l^4}{a^2 + l^2}\right)\dot{\theta} = 0$$

In the diagram (Fig. 1), P_0 is the initial position of P.

To obtain the kinetic energy of the system we should note that A is not a fixed point. Hence the rod is not simply rotating about A but has also kinetic energy of translation. It is therefore desirable to consider instead motion of the centre of mass of the rod G, and rotation about G considered as a fixed point. The latter gives no trouble, but a determination of the (two-dimensional) motion can.

A simple technique will be given here which it is not always necessary to adopt but which reduces to a merely mechanical process the calculation of translational kinetic energy in systems like the present one. We simply find the coordinates of G with respect to a suitable *fixed* origin, in this case P_0, and immediately obtain the velocity components of G, after which obtaining the kinetic energy is trivial.

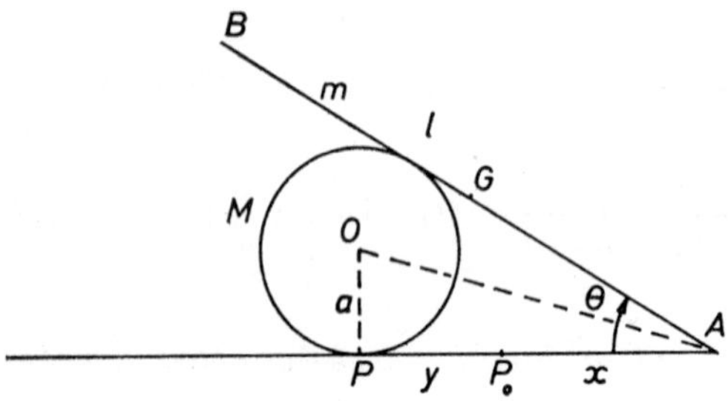

FIG. 1

Thus, the coordinates of G are $(x-\frac{1}{2}l\cos\theta, \frac{1}{2}l\sin\theta)$. The velocity components are therefore $(\dot{x}+\frac{1}{2}l\sin\theta\dot{\theta}, \frac{1}{2}l\cos\theta\dot{\theta})$. Hence,

$$T = \tfrac{1}{2}m[(\dot{x}+\tfrac{1}{2}l\sin\theta\dot{\theta})^2+(\tfrac{1}{2}l\cos\theta\dot{\theta})^2]+\tfrac{1}{24}ml^2\dot{\theta}^2+\tfrac{3}{4}M\dot{y}^2.$$

The last term on the right is of course the kinetic energy of the cylinder taking into account the fact that there is no slip between cylinder and table, so that the angular velocity of the cylinder is $\dot{y}/a$. Hence

$$T = \tfrac{3}{4}M\dot{y}^2+\tfrac{1}{2}m(\dot{x}^2+\dot{x}\dot{\theta}l\sin\theta)+\tfrac{1}{6}ml^2\dot{\theta}^2.$$

For the second part of the problem it is important to check on the number of degrees of freedom of the system and to ensure that this number equals the number of generalised coordinates chosen to define the dynamical state of the system. It is clear that, once x and y, say, are specified, so is θ. In fact,

$$\cot\tfrac{1}{2}\theta = (x+y)/a.$$

Thus the system has only two degrees of freedom, and must be described by two, and no more than two, generalised coordinates. Since $\dot{y}$ and $\dot{\theta}$ appear in the required relation, it is obvious that $\dot{x}$ must be eliminated from the expression so far obtained for the kinetic energy T. Further, the desired relationship between $\dot{y}$ and $\dot{\theta}$ must be an expression for the conservation of linear momentum of the system. Clearly, the potential energy will be independent of y, and the conservation law must be

$$\frac{\partial T}{\partial \dot{y}} = \text{constant}$$

In forming $\partial T/\partial \dot{y}$, we shall have $(\partial \dot{x}/\partial \dot{y})_{\theta,\dot{\theta}}$. From the geometrical relationship between x, y and θ, we obtain

$$\left(\frac{\partial \dot{x}}{\partial \dot{y}}\right)_{\theta,\dot{\theta}} = \left(\frac{\partial x}{\partial y}\right)_{\theta} = -1,$$

and

$$\frac{\partial T}{\partial \dot{y}} = \tfrac{3}{2}M\dot{y} - m\dot{x} - \tfrac{1}{2}m\dot{\theta}l \sin\theta,$$

$$= \text{constant} = 0,$$

since $\dot{x} = \dot{y} = \dot{\theta} = 0$, initially. Substituting

$$\dot{x} = -\tfrac{1}{2}a \operatorname{cosec}^2 \tfrac{1}{2}\theta \,.\, \dot{\theta} - \dot{y},$$

we find

$$(2m+3M)\dot{y} + m\dot{\theta}(a \operatorname{cosec}^2 \tfrac{1}{2}\theta - l \sin\theta) = 0.$$

This relation holds for all θ, provided the rod touches the cylinder tangentially. When the rod has slipped to the position where B just touches the cylinder, $\tan \frac{1}{2}\theta = a/l$. Hence at this point

$$(2m+3M)\dot{y} + \frac{m\dot{\theta}}{a}\left[a^2(1+l^2/a^2) - \frac{l\,.\,a\,.\,2a/l}{1+a^2/l^2}\right] = 0,$$

which simplifies to the required relation

$$(2m+3M)\dot{y} + \frac{m}{a}\dot{\theta}\left(\frac{a^4+l^4}{a^2+l^2}\right) = 0.$$

1.2. *Two smooth wedges of mass M, with right-angled isosceles triangular cross-sections, rest with their hypotenuses in contact with a horizontal plane. A uniform rod of length 2a and mass m is placed gently to touch each wedge in the plane of symmetry, the rod making an angle α with the horizontal. The system is then released from rest.*

If the centre of gravity of the rod is at a height y above the plane, and has moved through a horizontal distance x when the inclination of the rod to the horizontal is θ, find the kinetic energy, and hence prove that

$$(8M+m)a\dot{\theta}^2 = 12Mg(\cos\alpha-\cos\theta).$$

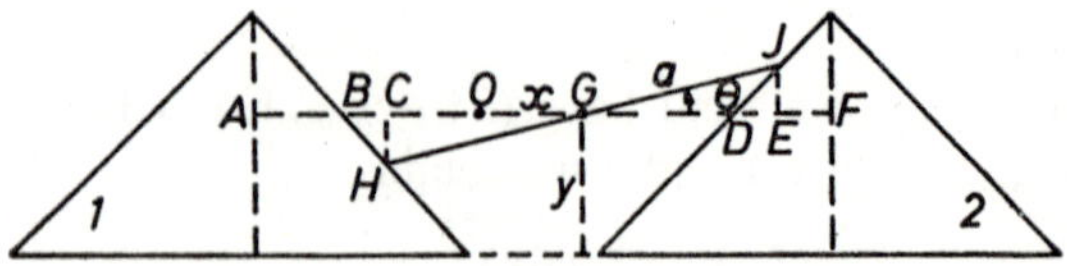

FIG. 2

$$T = \text{kinetic energy of rod } (T_R)+\text{kinetic energy of wedge 1 } (T_1) + \text{kinetic energy of wedge 2 } (T_2)$$
$$= T_R+T_1+T_2.$$
$$T_R = \tfrac{1}{2}m(\dot{x}^2+\dot{y}^2)+\tfrac{1}{6}ma^2\dot{\theta}^2.$$

Let the wedges be of height l.

$$AG = CG+BC+AB = CG+CH+(l-y).$$
$$AO = CG+CH+(l-y)-x,$$
$$= a(\cos\theta+\sin\theta)+l-x-y.$$

Hence

$$T_1 = \tfrac{1}{2}M[a(\cos\theta-\sin\theta)\dot{\theta}-\dot{x}-\dot{y}]^2.$$

Similarly,

$$T_2 = \tfrac{1}{2}M[-a(\sin\theta+\cos\theta)\dot{\theta}+\dot{x}-\dot{y}]^2,$$

and

$$T_1+T_2 = M[a^2\dot{\theta}^2+\dot{x}^2+\dot{y}^2-2a\dot{\theta}(\dot{x}\cos\theta-\dot{y}\sin\theta)].$$

Potential energy is given by $V = mgy$, and the Lagrangian of the system by

$$L = T_R+T_1+T_2-V.$$

To specify the position of the rod requires three coordinates, the coordinates of $G(x, y)$ and the angle θ the rod makes with the horizontal. Once the position of the rod is specified the positions of the wedges are also fixed, as we have seen in the foregoing analysis.

Thus the whole system is a holonomic system of three degrees of freedom, and it is valid to obtain Lagrange's equations for x, y and θ. These are, respectively,

$$\frac{d}{dt}[m\dot{x}+2M(\dot{x}-a\dot{\theta}\cos\theta)] = 0. \tag{1}$$

$$\frac{d}{dt}[m\dot{y}+2M(\dot{y}+a\dot{\theta}\sin\theta)] = -mg. \tag{2}$$

$$\frac{d}{dt}[(\tfrac{1}{3}m+2M)a\dot{\theta}-2M(\dot{x}\cos\theta-\dot{y}\sin\theta)] = 2M\dot{\theta}(\dot{x}\sin\theta+\dot{y}\cos\theta). \tag{3}$$

Equation (*1*) expresses the conservation of linear momentum of the whole system in the x-direction. From (*3*), we obtain

$$(\tfrac{1}{3}m+2M)a\ddot{\theta} = 2M(\ddot{x}\cos\theta-\ddot{y}\sin\theta). \tag{4}$$

Equations (*1*) and (*2*) yield expressions for $\ddot{x}$ and $\ddot{y}$, namely,

$$\ddot{x} = \frac{2Ma}{m+2M}(\ddot{\theta}\cos\theta-\dot{\theta}^2\sin\theta).$$

$$\ddot{y} = \frac{-mg}{m+2M}-\frac{2Ma}{m+2M}(\ddot{\theta}\sin\theta+\dot{\theta}^2\cos\theta).$$

Hence,

$$\ddot{x}\cos\theta-\ddot{y}\sin\theta = \frac{2Ma}{m+2M}\ddot{\theta}+\frac{mg\sin\theta}{m+2M}.$$

Inserting this expression into equation (*4*) leads to

$$(\tfrac{1}{3}m+2M)a\ddot{\theta} = \frac{2M}{m+2M}(2Ma\ddot{\theta}+mg\sin\theta),$$

which simplifies to the form

$$(m+8M)a\ddot{\theta} = 6Mg\sin\theta.$$

Integrating this equation with respect to θ and setting $\theta = \alpha$ initially, we obtain

$$(m+8M)a\dot{\theta}^2 = 12Mg(\cos\alpha-\cos\theta).$$

1.3. *Show that the kinetic energy of an axisymmetric rigid body, rotating freely about its mass centre, which is fixed, can be expressed in the form*

$$T = \tfrac{1}{2}I(\dot{\theta}^2+\dot{\phi}^2 \sin^2\theta)+\tfrac{1}{2}I_3(\dot{\phi}\cos\theta+\dot{\psi})^2$$

Show that (i) $\dot{\phi}\cos\theta+\dot{\psi}$, (ii) $I\dot{\phi}\sin^2\theta+I_3\cos\theta(\dot{\phi}\cos\theta+\dot{\psi})$ *are constants of the motion, and discuss briefly their physical meaning. State any other constant of the motion.*

Derive an equation, in the variable $u = \cos\theta$, *whose solutions give the values of* θ *at which* $\dot{\theta}$ *vanishes. Verify that, when the direction* $\theta = 0$ *is chosen along the total angular momentum of the body, this equation has equal roots. Discuss the physical meaning of this result.*

The first part of the question deals with the well known form of the kinetic energy of rotation of a rigid body, expressed in terms of the Euler angles θ, ϕ, ψ. A derivation will not be given here, but the student should quickly check that he understands the principles in obtaining the result.

The Euler angles are valid generalised coordinates for rigid body rotation. Since in this case we are considering rotation about the (fixed) mass centre, there is no potential energy term. Then, since T is independent of ϕ and ψ, we immediately obtain two constants of the motion:

(*i*) $\dfrac{\partial T}{\partial \dot{\psi}}$ = constant; hence $\dot{\phi}\cos\theta+\dot{\psi}$ = constant

(*ii*) $\dfrac{\partial T}{\partial \dot{\phi}}$ = constant; hence $I\dot{\phi}\sin^2\theta+I_3\cos\theta(\dot{\phi}\cos\theta+\dot{\psi})$ = constant

(*i*) expresses conservation of the component of angular momentum (or angular velocity) about the symmetry axis. (*ii*) expresses conservation of the component of angular momentum about the polar axis of the Euler angles, namely the direction $\theta = 0$. A third constant of the motion is obviously the total kinetic energy, $T = E$.

Let $\dot{\phi}\cos\theta+\dot{\psi} = \zeta$, $I\dot{\phi}\sin^2\theta+I_3\zeta\cos\theta = A$. The kinetic energy can be written in terms of θ, $\dot{\theta}$ alone, eliminating $\dot{\phi}$ and $\dot{\psi}$ in terms of A and ζ. We obtain

$$\tfrac{1}{2}I_3\zeta^2+\tfrac{1}{2}I\dot{\theta}^2+\tfrac{1}{2}I\sin^2\theta\left(\frac{A-I_3\zeta\cos\theta}{I\sin^2\theta}\right)^2 = E.$$

Hence,

$$\dot{\theta}^2 \sin^2 \theta = \frac{2E - I_3\zeta^2}{I} \sin^2 \theta - \left(\frac{A}{I} - \frac{I_3\zeta}{I} \cos \theta\right)^2.$$

Let $u = \cos \theta$, $\alpha = (2E - I_3\zeta^2)/I$, $\beta = A/I$, $\gamma = I_3\zeta/I$. Then

$$\dot{u}^2 = (1-u^2)\alpha - (\beta - \gamma u)^2 \equiv f(u).$$

We are interested in the roots of the equation $f(u) = 0$; if these roots are ≤ 1 in magnitude, they will correspond to a physically realizable value for θ at which the nutation angular velocity $\dot{\theta}$ vanishes. From a purely mathematical point of view, it is more convenient to extend the range of u to all real values, not just $|u| \leq 1$. The latter condition will define the 'physical region' in u. It is also possible mathematically for $f(u)$ to become negative, but this implies that $\dot{u}^2$ is negative, again physically impossible. Thus a diagram of $f(u)$ against u varying from $-\infty$ to ∞ will have a physical region defined by $|u| \leq 1$ and $f \geq 0$. These comments are generally applicable. In this problem the roots are given by

$$(1-u^2)\alpha - (\beta - \gamma u)^2 = 0,$$

i.e.

$$(\alpha + \gamma^2)u^2 - 2\beta\gamma u + \beta^2 - \alpha = 0.$$

Equal roots will occur if

$$\beta^2\gamma^2 = (\gamma^2 + \alpha)(\beta^2 - \alpha)$$

or

$$\alpha(\beta^2 - \gamma^2) = \alpha^2.$$

Assuming $\alpha \neq 0$, $\beta^2 - \gamma^2 = \alpha$. We now *choose* the direction $\theta = 0$ along the total angular momentum $\boldsymbol{L}$. As a consequence, we can evaluate A and E, namely,

$$A = \boldsymbol{L}$$

$$E = \tfrac{1}{2}I_3\zeta^2 + \tfrac{1}{2}I\eta^2, \qquad (1)$$

where η is the component of angular velocity perpendicular to the symmetry axis.

$$A^2 = \boldsymbol{L}^2 = I^2\eta^2 + I_3^2\zeta^2 \qquad (2)$$

Eliminating η from equations (*1*) and (*2*), we have

$$\frac{A^2 - I_3^2\zeta^2}{I^2} = \frac{2E - I_3\zeta^2}{I},$$

i.e.

$$\beta^2 - \gamma^2 = \alpha.$$

This verifies that, when $\theta = 0$ direction is chosen along $\boldsymbol{L}$, the equation $f(u) = 0$ has equal roots.

Mathematically, the curve $f(u)$ is tangential to the u-axis. Since $f(u) < 0$ as $|u| \to \infty$, the curve lies entirely below the u-axis except at the point of contact, u_0, where

$$u_0 = \beta\gamma/(\alpha+\gamma^2) = \gamma/\beta.$$

By substituting for β and γ in terms of ζ and $\boldsymbol{L}$, one can easily show that $|u_0| = |I_3\zeta|/L \leqq 1$.

Physically, for this choice of axes, $\dot{\theta}$ vanishes at a point $\theta_0 = \cos^{-1} u_0$, and no neighbouring motion is possible. The motion of the body is stable at $\theta = \theta_0$. Precession about the axis $\theta = 0$ will take place, with constant precessional angular velocity.

These results are usually obtained by considering Euler's equations, but the method illustrated here is of importance in dynamics. For example, one obtains considerable insight into the motion of a symmetrical top under gravity, an exact solution of which involves elliptic functions.

1.4. *A uniform rectangular lamina, whose principal moments of inertia are I, kI and* $(1+k)I$, *is mounted so that it can swing freely about a smooth horizontal axis which coincides with the axis about which the moment of inertia is I. The system is carried on a light frame which can rotate about a smooth vertical axis passing through the centre of the lamina. At time* $t = 0$, $\theta = \pi/4$, $\dot{\theta} = 0$, $\dot{\phi} = \Omega$, *where* θ *is the inclination of the lamina to the horizontal and* ϕ *is the angle through which the frame turns. Show that, in the subsequent motion.*

$$\dot{\theta}^2 = \frac{(1+2k)\Omega^2 \cos 2\theta}{4(k+\cos^2\theta)}.$$

The angular velocity $\boldsymbol{\omega}$ of the lamina is given by

$$\boldsymbol{\omega} = \dot{\theta}\mathbf{i} + \dot{\phi}\mathbf{z}$$

(an infinitesimal rotation $\delta\theta$ is about **i**, $\delta\phi$ is about **z**). Hence in terms of unit vectors **i**, **j**, **k** as a set of rotating axes, $\boldsymbol{\omega}$ has components $(\dot{\theta}, \dot{\phi} \sin \theta, \dot{\phi} \cos \theta)$. Since no work is done on the lamina by external forces (the axis of rotation is smooth), and the mass centre of the lamina is fixed, its kinetic energy is conserved. Since no torque can act about the vertical axis about which the whole system rotates, the vertical component of angular momentum of the system is conserved. To make use of these two conservation laws, let us consider θ and ϕ as generalised coordinates of the system. Then

$$T = \tfrac{1}{2}I\dot{\theta}^2 + \tfrac{1}{2}kI\dot{\phi}^2 \sin^2 \theta + \tfrac{1}{2}(1+k)I\dot{\phi}^2 \cos^2 \theta,$$
$$= \tfrac{1}{2}I\dot{\theta}^2 + \tfrac{1}{2}(k+\cos^2 \theta)I\dot{\phi}^2.$$

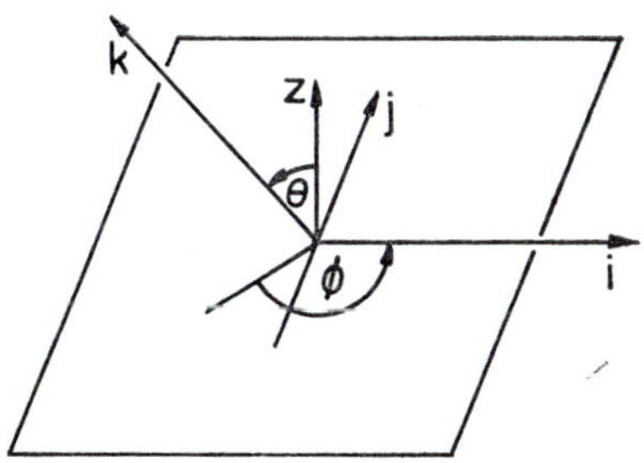

FIG. 3

The conservation of angular momentum about the vertical axis of rotation has expression in the fact that T is independent of ϕ. As usual, this means that $\partial T/\partial \dot{\phi}$ is a constant of the motion,

i.e.

$$(k+\cos^2 \theta)\dot{\phi} = \text{constant}$$
$$= (k+\tfrac{1}{2})\Omega, \qquad (1)$$

since initially $\theta = \pi/4$ and $\dot{\phi} = \Omega$.
T is also a constant of the motion and so

$$\dot{\theta}^2 + (k+\cos^2 \theta)\dot{\phi}^2 = (k+\tfrac{1}{2})\Omega^2, \qquad (2)$$

where now we have also used the fact that $\dot{\theta}$ initially vanishes.

The relations (*1*) and (*2*) enable us to eliminate $\dot{\phi}$ and obtain

$$\dot{\theta}^2 + (k+\cos^2 \theta)\left[\frac{(k+\frac{1}{2})\Omega}{k+\cos^2 \theta}\right]^2 = (k+\tfrac{1}{2})\Omega^2,$$

which simplifies to the required condition

$$\dot{\theta}^2 = \frac{(2k+1)\Omega^2 \cos 2\theta}{4(k+\cos^2\theta)}.$$

1.5. *To a fixed point O are smoothly jointed two uniform rods of length 2a and mass m, and on them slides, by means of a smooth ring at each end, a third rod identical with the first two. Initially the three rods are in a horizontal line with the third rod below the first two and its ends at their mid points. The whole system rotates in a horizontal plane about O with angular velocity ω. The system is then released. Show that the third rod will slide off unless* $\omega^2 > 2g/a\sqrt{3}$. *(It should be assumed that the third rod remains horizontal throughout.)*

It should be immediately obvious that the method of Lagrange's equations for generalised coordinates is applicable here. It is best in such problems to consider the physics first before setting up any mathematical equations. Firstly there are two degrees of freedom and so we must choose *two* suitable coordinates as generalised coordinates. One must always try and make use of any symmetry property, and here we have symmetry about the vertical axis through O. Thus we shall find it useful to define one of the generalised coordinates, say ϕ, to be the angle through which the system of rods has turned about this axis. Since the system is torque-free about the vertical axis, we may anticipate that the vertical component of the total angular momentum is conserved. The second coordinate we take to be θ, the angle between the first rod and the horizontal. By symmetry, this will also be the angle between the second rod and the horizontal.

Initial conditions are $\theta = \dot{\theta} = 0$; $\phi = 0$, $\dot{\phi} = \omega$. Referring to Fig. 4, we have $OM = a\tan\theta$, velocity of $M = a\sec^2\theta\dot{\theta}$. We require expressions for the kinetic and potential energies, T and V respectively, of the system.

$$T = \tfrac{1}{6}ma^2(3\sec^4\theta\dot{\theta}^2+\dot{\phi}^2)+\tfrac{4}{3}ma^2(\dot{\theta}^2+\cos^2\theta\dot{\phi}^2).$$

$$V = -mga\tan\theta-2mga\sin\theta.$$

In T, the first term gives the kinetic energy of the third rod, which remains horizontal throughout the motion; the second term gives the kinetic energy of the other two rods (remember the factor 2).

At this point we could proceed by writing down the two Lagrange equations for θ and ϕ. However, we have already noted that the vertical component of angular momentum is conserved. This is now seen to be

$$\frac{\partial T}{\partial \dot{\phi}} = \text{constant} \tag{1}$$

since ϕ is absent from T and V (ϕ is an ignorable coordinate).

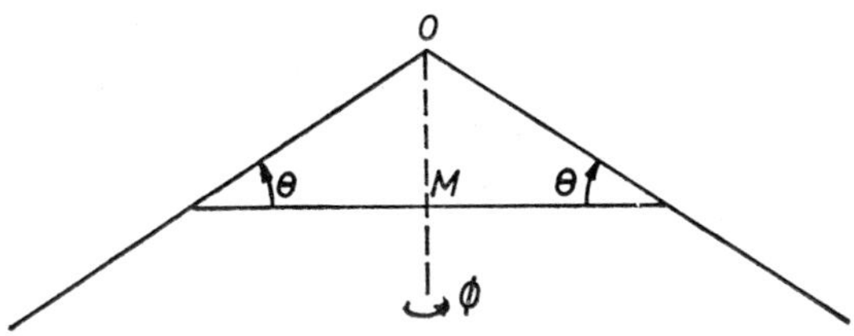

FIG. 4

A second constant of the motion is obviously the total energy,

$$T+V = \text{constant}. \tag{2}$$

The student can quickly verify that this is a first integral of Lagrange's equation for θ. Thus we obtain from (*1*)

$$(\tfrac{1}{6}+\tfrac{4}{3}\cos^2\theta)\dot{\phi} = \text{constant}, \tag{3}$$

i.e., $\frac{1}{6}(1+8\cos^2\theta)\dot{\phi} = \frac{3}{2}\omega$, using the initial conditions. From (*2*), we have

$$A\dot{\theta}^2+\tfrac{1}{6}\frac{81\omega^2}{1+8\cos^2\theta}-\frac{g}{a}(\tan\theta+2\sin\theta)$$
$$= \text{constant}$$
$$= \tfrac{3}{2}\omega^2, \text{ again using initial conditions.} \tag{4}$$

In this energy equation, the actual form of A does not matter, except that it is positive definite. Note that $\dot{\phi}$ has been eliminated, using (*3*). What we have obtained in (*4*) is an equation of the form

$$\dot{\theta}^2 = f(\theta),$$

into which initial conditions have been incorporated. It is useful at this stage in a problem to make any simple check that all is well. Thus, $\dot{\theta} = 0$ when $\theta = 0$, i.e. $f(0) = 0$.

We are now in a position to evaluate $\dot{\theta}^2$ at any later state of the system, defined by θ. If, for any value of θ, $f(\theta)$ once more vanishes (it must clearly first become positive) this will define a turning point for the motion, $\dot{\theta}$ will change sign and the rods will begin to rise again. We have now to consider whether this will occur before the third rod falls off. That is, we now have a geometrical condition which will lead to the required condition on ω. Obviously, in the critical position the rods form an equilateral triangle, with $\theta = 60°$. The condition that the rod does not slide off is that $f(\theta)$ has a zero for $\theta < 60°$, or $f(\theta) < 0$, $\theta = 60°$. Inserting the various trigonometric functions evaluated at 60°, we obtain

$$\omega^2 > 2g/a\sqrt{3}.$$

1.6. *A hollow sphere S_1, of radius b, has another hollow sphere S_2, of radius a, free to roll inside it. Initially the sphere S_1 is at rest on a horizontal plane and the sphere S_2 is at rest in its equilibrium position. The sphere S_1 is then seized and forced to slide along the plane, without rotation, with constant velocity v. If, in the subsequent motion, θ denotes the angle between the line joining the centres of the two spheres and the vertical show that*

$$\dot{\theta}^2 = \frac{9v^2}{25c^2} - \frac{6g}{5c}(1-\cos\theta), \quad \text{where } c = b-a.$$

Show that the inner sphere will perform complete revolutions if $9v^2 > 85gc$.

The energy equation, namely, kinetic+potential energy = constant, is valid only when constraints in the system are independent of time. We cite as an example rigid body constraints. However when the constraints are time-dependent the result quoted above no longer applies. Provided that the Lagrangian of the system is still independent of time, a simple modification of the energy equation is possible. More precisely, if we have time-independent constraints

$$T+V = \text{constant},$$

where V is the potential energy, and the kinetic energy T is in general a quadratic form of the generalised velocities. We express this fact by setting $T = T_2$. When the constraints depend on time explicitly, the kinetic energy contains terms linear in and independent of the generalised velocities, namely, $T = T_2+T_1+T_0$.

Although the constraints are now time-dependent, L may still be independent of time (except through the time dependence of the generalised coordinates and velocities). In this case $T+V$ is no longer conserved. The energy conservation equation is replaced by

$$T_2-T_0+V = \text{constant}.$$

[To prove this result, form

$$H = \sum_i \dot{q}_i \frac{\partial L}{\partial \dot{q}_i} - L$$

and show that $dH/dt = 0$ provided $\partial L/\partial t = 0$. Then, using Euler's theorem on homogeneous functions, it follows that

$$H = 2T_2+T_1-L = T_2-T_0+V.$$

H is the Hamiltonian of the system, still expressed in terms of generalised coordinates and velocities.]

The discussion above is given since the result does not appear to be too well known, although it is mentioned in some textbooks.

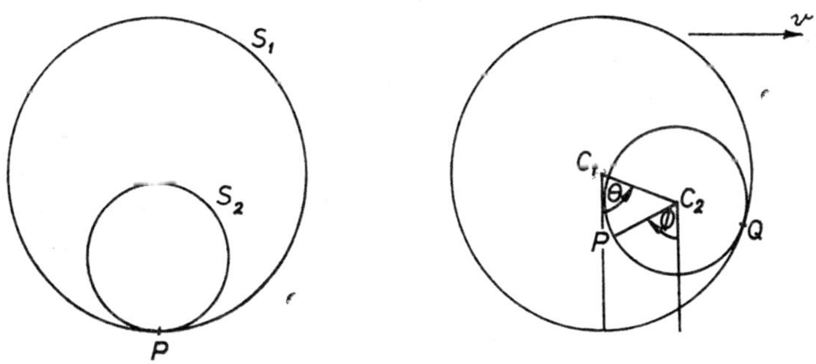

FIG. 5

T = kinetic energy of sphere S_2

= kinetic energy of translation + kinetic energy of rotation about the centre of mass C_2

$= \frac{1}{2}mw^2 + \frac{1}{2} \cdot \frac{2}{3}ma^2\dot{\phi}^2.$

Taking components of $\mathbf{w}$, the velocity of C_1, along and perpendicular to the plane, we obtain

$$\mathbf{w} = (v+(b-a)\dot{\theta} \cos \theta, (b-a)\dot{\theta} \sin \theta).$$

Hence

$$\mathbf{w}^2 = v^2+2v(b-a)\dot{\theta} \cos \theta+(b-a)^2\dot{\theta}^2.$$

Note the appearance in $\mathbf{w}^2$ of terms linear in $\dot{\theta}$, and independent of $\dot{\theta}$. We need a relationship between $\dot{\phi}$ and $\dot{\theta}$. Relative to the large sphere, Q, the point of contact of the two spheres, is stationary. Relative to C_2, Q has velocity $a\dot{\phi}$; relative to C_1, its velocity is $a\dot{\phi}+(b-a)\dot{\theta}$, and this must be zero.

Hence,

$$\begin{aligned} T &= \tfrac{1}{2}m[v^2+2v(b-a)\dot{\theta} \cos \theta+(b-a)^2\dot{\theta}^2]+\tfrac{1}{3}m(b-a)^2\dot{\theta}^2 \\ &= T_2+T_1+T_0 \\ V &= mg(b-a)(1-\cos \theta) \end{aligned}$$

Since $T_0 = \tfrac{1}{2}mv^2 = \text{constant}$, we may write

$$\tfrac{5}{6}m(b-a)^2\dot{\theta}^2-mg(b-a) \cos \theta = \text{constant},$$

i.e.,

$$\dot{\theta}^2 = K-\tfrac{6}{5}(g/c)(1-\cos \theta), \quad c = b-a.$$

Now we have to determine the constant K from consideration of initial conditions. Initially, an impulse acts on the large sphere and is transmitted to the small sphere through the point of contact P. P as a point on the large sphere has velocity v; hence as a point on the small sphere its velocity is also v, since there is no slip. Suppose C_2 has initially velocity u, and that an impulse I acts on the small sphere at P. We can write two equations involving the change in linear and angular momentum due to the action of the impulse.

$$I = mu.$$

$$Ia = \tfrac{2}{3}ma^2 \,.\, (v-u)/a.$$

i.e.

$$I = \tfrac{2}{3}m(v-u).$$

Hence

$$u = \tfrac{2}{3}(v-u), \quad \text{or} \quad v-u = \tfrac{3}{5}v.$$

Initially,

$$\theta = 0 \quad \text{and} \quad \dot{\theta} = (v-u)/c = \tfrac{3}{5}v/c,$$

so that

$$K = \tfrac{9}{25}(v/c)^2.$$

The small sphere will perform complete revolutions if the reaction of S_1 on S_2 is positive when $\theta = \pi$. The condition is therefore

$$m\dot{\theta}^2 c > mg.$$

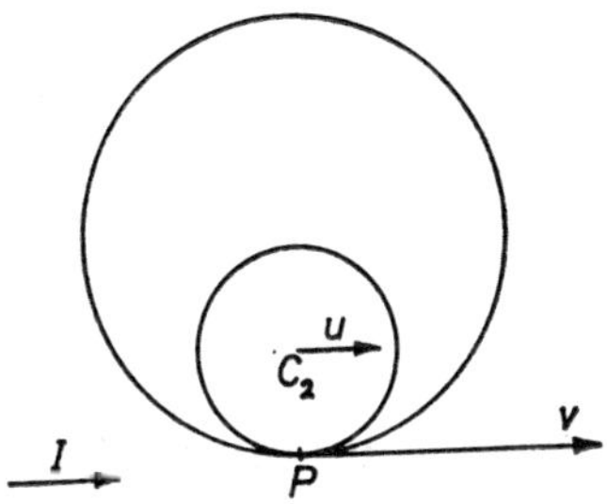

FIG. 6

It is easily seen by substitution that this condition is

$$v^2 > 85gc/9.$$

Impulsive motion

For a single particle, if a force **F** acts on it for a time δt, not necessarily short, the change in momentum of the particle is obtained by integrating the equation of motion from $t = 0$ to $t = \delta t$.

$$\delta \mathbf{p} = \int_0^{\delta t} \frac{d\mathbf{p}}{dt}\, dt = \int_0^{\delta t} \mathbf{F}\, dt = \mathbf{I}$$
$$= \text{impulse of } \mathbf{F}.$$

Impulsive forces are taken to be forces which act for a very much shorter time than other forces, so that we can usually neglect the effect of the latter in impulsive motion, and also any change in position of the particle. In such cases, **I** is defined to be the limit

$$\mathbf{I} = \mathscr{L}_{\delta t \to 0} \int_0^{\delta t} \mathbf{F}\, dt.$$

For a system of particles, acted on by a set of forces $\mathbf{F}_i$, $\mathbf{F}_i$ being the force on the ith particle, the total linear momentum $\mathbf{P} = \sum_i m_i\dot{\mathbf{r}}_i$. For impulsive forces

$$\delta\mathbf{P} = \sum_i \mathop{\mathscr{L}}_{\delta t\to 0} \int_0^{\delta t} F_i\,dt = \sum_i \mathbf{I}_i.$$

The total angular momentum $\mathbf{L}$ with respect to a fixed origin O satisfies the equation

$$\frac{d\mathbf{L}}{dt} = \mathbf{N} = \sum_i \mathbf{r}_i \times \mathbf{F}_i$$

Hence

$$\delta\mathbf{L} = \sum_i \mathop{\mathscr{L}}_{\delta t\to 0} \int_0^{\delta t} \mathbf{r}_i \times \mathbf{F}_i\,dt$$

Recalling that we may neglect any change in $\mathbf{r}_i$, we obtain

$$\delta\mathbf{L} = \sum_i \mathbf{r}_i \times \mathbf{I}_i$$

These equations for $\delta\mathbf{P}$ and $\delta\mathbf{L}$ can be applied to rigid body problems. Note too that the origin O need not even be fixed, since it is effectively fixed during the application of the impulsive forces. In many problems, however, the dynamical system in question is best described in terms of generalised coordinates and Lagrange's equations. We therefore sketch the counterpart of Lagrange's equations for impulsive forces.

Using the usual notation, Lagrange's equations are of the form

$$\frac{d}{dt}\left(\frac{\partial T}{\partial \dot{q}_i}\right) - \frac{\partial T}{\partial q_i} = Q_i,$$

where T is the kinetic energy function, q_i the ith generalised coordinate, and Q_i the corresponding generalised force. Integrating over a short time δt we obtain the generalised impulse J_i:

$$J_i = \mathop{\mathscr{L}}_{\delta t\to 0} \int_0^{\delta t} Q_i\,dt = \mathop{\mathscr{L}}_{\delta t\to 0} \int_0^{\delta t} dt\left[\frac{d}{dt}\left(\frac{\partial T}{\partial \dot{q}_i}\right) - \frac{\partial T}{\partial q_i}\right].$$

Now, $\partial T/\partial q_i$ is a known function of q_i, $\dot{q}_i$ (all i) and is finite. Hence this term vanishes as $\delta t \to 0$. We thus have

$$J_i = \delta\left(\frac{\partial T}{\partial \dot{q}_i}\right).$$

Since the q_i have fixed values while the impulse acts, only the generalised velocities $\dot{q}_i$ need to be considered. Let $\dot{q}_i = u_i$ = generalised velocity. Then

$$J_i = \delta\left(\frac{\partial T}{\partial u_i}\right).$$

Recall that Q_j is defined by the relation

$$\sum_i \mathbf{F}_i \cdot \delta\mathbf{r}_i = \sum_j Q_j\, \delta q_j.$$

Thus,

$$\sum_i \mathbf{I}_i \cdot \delta\mathbf{r}_i = \sum_j J_j\, \delta q_j.$$

We can use this relation to obtain a more convenient definition of the generalised impulse. Let $\mathbf{v}_i$ be the velocity of the ith particle of the system in a motion consistent with the constraints in the system. Then, for any $\delta\mathbf{r}_i$

$$\sum_i \mathbf{I}_i \cdot \frac{\delta\mathbf{r}_i}{\delta t} = \sum_j J_j \frac{\delta q_j}{\delta t},$$

i.e.

$$\sum_i \mathbf{I}_i \cdot \mathbf{v}_i = \sum_j J_j\, u_j.$$

To obtain J_j, we have to express $\mathbf{v}_i$ in terms of the u_j, and then equate coefficients of u_j on both sides of the equation.

Before discussing more involved problems on impulsive motion, the foregoing discussion will be made clearer by means of a simple example.

A rod of length 2a, mass m, lying on a smooth horizontal table, is struck a blow of impulse I at one end perpendicular to the rod. Discuss the motion.

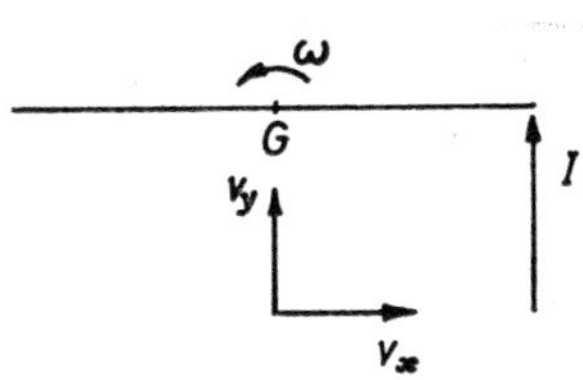

FIG. 7

(*i*) *Conventional method*

After the impulse acts, let the centre of mass G have velocity components v_x and v_y and let the angular velocity of the rod be ω.

Linear momentum: $mv_x = 0$, $mv_y = I$. Hence $v_x = 0$, $v_y = I/m$.
Angular momentum: $\frac{1}{3}ma^2\omega = Ia$, so that $\omega = 3I/ma$.

(*ii*) *Method using generalised coordinates*

$v_y = u$, and ω are a suitable set of generalised velocities.

$$T = \tfrac{1}{2}mu^2 + \tfrac{1}{6}ma^2\omega^2.$$

Then,

$$J_u u + J_\omega \omega = I(u + a\omega).$$

On the right, $u + a\omega$ = velocity of point of application of I, in terms of u and ω.

Equating coefficients, we obtain

$$J_u = I, \quad J_\omega = Ia.$$

We have now to use the equations of impulsive motion:

$$\delta\left(\frac{\partial T}{\partial u}\right) = J_u, \text{ hence } mu = I, \text{ and } u = I/m.$$

$$\delta\left(\frac{\partial T}{\partial \omega}\right) = J_\omega, \text{ hence } \tfrac{1}{3}ma^2\omega = Ia, \text{ and } \omega = 3I/ma.$$

Although in this example there is more work involved in obtaining the solution using generalised velocities, it is included to illustrate the method. In the case of more involved problems, such as complicated frameworks subjected to impulsive forces, the method is very powerful, as we shall see.

1.7. *A square framework ABCD, made up of four equal rods of mass m and length 2a, smoothly jointed together, lies on a smooth horizontal table. It is acted upon by a blow of impulse I at the vertex A in the direction of the diagonal AC. Determine the initial state of motion.*

We choose as generalised velocities

V = velocity of A.

ω = magnitude of angular velocity of each rod.
(by symmetry these are all equal)

Let P, Q, R, S be the mass centres of the rods, as shown in the diagram. We require the kinetic energy T of the framework, and therefore the velocities of A, B, C, and D from which we can find the velocities of P, Q, R and S. We make use of the line of symmetry AC.

$$\begin{aligned} \text{Velocity of } A &= (V, O) \\ \text{velocity of } B &= (V-\sqrt{2}\, a\omega, -\sqrt{2}\, a\omega) \\ \text{velocity of } C &= (V-2\sqrt{2}\, a\omega, 0) \end{aligned}$$

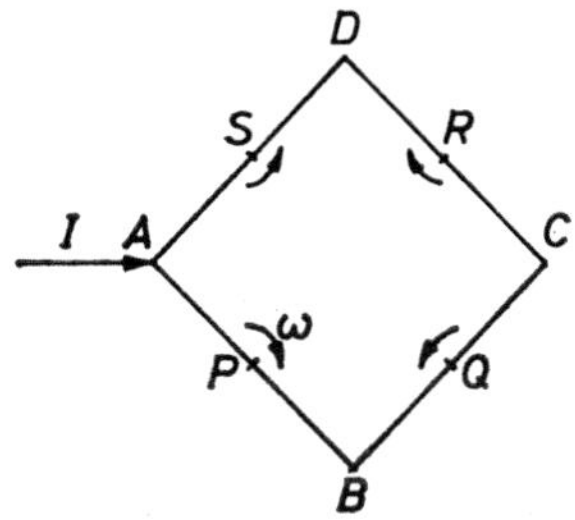

FIG. 8

To obtain these results, we take account of the fact that, by symmetry, A and C will move only along the initial direction of AC, (the x-axis). Further, relative to A, B moves instantaneously in a circle with angular velocity ω, in the direction CB. Its velocity is therefore $2a\omega$. Resolving in the x- and y-directions gives the result above. For the velocity of C we note that

$$v_{Bx} = \tfrac{1}{2}(v_{Ax}+v_{Cx}).$$

We can easily obtain the velocity components of P and Q, the mid-points of AB and BC.

$$\begin{aligned} \text{Velocity of } P &= (V-a\omega/\sqrt{2}, -a\omega/\sqrt{2}) \\ \text{Velocity of } Q &= (V-3a\omega/\sqrt{2}, -a\omega/\sqrt{2}) \end{aligned}$$

Kinetic energy $T = 2T_{AB}+2T_{BC}$.

$$\begin{aligned} T_{AB} &= \tfrac{1}{2}m[(V-a\omega/\sqrt{2})^2+\tfrac{1}{2}a^2\omega^2+\tfrac{1}{3}a^2\omega^2]. \\ T_{BC} &= \tfrac{1}{2}m[(V-3a\omega/\sqrt{2})^2+\tfrac{1}{2}a^2\omega^2+\tfrac{1}{3}a^2\omega^2]. \end{aligned}$$

The last term in each case is the rotational kinetic energy. Hence

$$T/m = 2v^2 - 4\sqrt{2}\, a\omega V + 20a^2\omega^2/3.$$

Next we determine J_V and J_ω.

$$J_V V + J_\omega \omega = I \times \text{velocity of } A = IV.$$

Hence

$$J_V = I \quad \text{and} \quad J_\omega = 0.$$

Equations of impulsive motion are

$$\frac{\partial T}{\partial V} = I, \quad \frac{\partial T}{\partial \omega} = 0.$$

Thus,

$$4V - 4\sqrt{2}\, a\omega = I/m.$$

$$-4\sqrt{2}\, aV + 40a^2\omega/3 = 0.$$

giving

$$V = 5I/8m, \quad \omega = 3\sqrt{2}\, I/16ma.$$

1.8. *Two equal uniform rods AB and BC, freely jointed at B, lie on a smooth horizontal plane and ∠ABC is a right angle. If an impulse X is applied at A along AB, and an impulse Y in the direction BC, find the kinetic energy T of the rods, and show that if the same impulse is applied with C fixed, the kinetic energy is* $\frac{15}{16}T$.

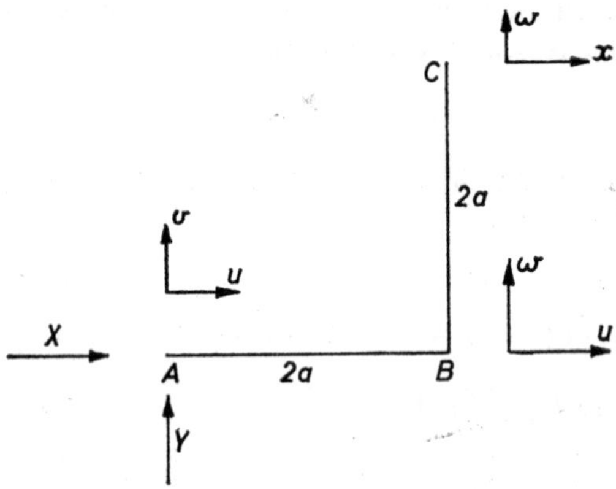

FIG. 9

As shown in the diagram (Fig. 9), let A have velocity components (u, v) along directions AB, BC. Let B have velocity components (u, w) and $C(x, w)$. We have taken account of the rigidity of the rods.

We first calculate the kinetic energy of rod AB. It is given by

$$T_{AB} = \tfrac{1}{2}m\left[u^2+\left(\frac{v+w}{2}\right)^2\right]+\tfrac{1}{6}ma^2\left(\frac{w-v}{2a}\right)^2$$

$$= \tfrac{1}{6}m(3u^2+v^2+w^2+vw).$$

In the same way,

$$T_{BC} = \tfrac{1}{6}m(3w^2+u^2+x^2+ux),$$

so that

$$T = \tfrac{1}{6}m(4u^2+v^2+4w^2+x^2+vw+ux).$$

The generalised impulses are obtained from the relation

$$J_u u+J_v v+J_w w+J_x x = Xu+Yv.$$

Thus,

$$J_u = X, \quad J_v = Y, \quad J_w = J_x = 0.$$

The impulsive equations of motion are $\partial T/\partial u = J_u$, $\partial T/\partial v = J_v$, and so on, giving

$$\tfrac{1}{6}m(8u+x) = X; \quad v+8w = 0;$$

$$\tfrac{1}{6}m(2v+w) = Y; \quad 2x+u = 0,$$

whence,

$$u = 4X/5m; \quad w = -2Y/5m;$$

$$v = 16Y/5m; \quad x = -2X/5m.$$

Substituting these into the expression for T, we obtain

$$T = 2(X^2+4Y^2)/5m.$$

If C is held fixed, $w = x = 0$, and the number of degrees of freedom is reduced from four to two. Only J_u and J_v occur, determined now by the equations

$$u = 3X/4m, \quad v = 3Y/m.$$

The kinetic energy is now

$$T' = \tfrac{1}{6}m(3u^2+v^2)$$
$$= 3(X^2+4Y^2)/8m.$$

i.e.

$$T' = \tfrac{15}{16}T.$$

EXERCISES

1. A simple pendulum consists of a mass attached by a light rod, length l, to its point of suspension O, which is forced to move in simple harmonic motion along a horizontal straight line with coordinate ξ given by $\xi = a \sin \omega t$. If, at time $t = 0$, the bob of the pendulum is at rest vertically below O show that after a complete period of the simple harmonic motion the horizontal displacement x of the bob is given by

$$x = -\frac{a\omega \sqrt{\dfrac{g}{l}} \sin \dfrac{2\pi}{\omega}\sqrt{\dfrac{g}{l}}}{\left(\dfrac{g}{l}-\omega^2\right)}.$$

It may be assumed that during the motion

$$|x-\xi| \ll l.$$

2. A long ruler of mass m_1 rests on a smooth horizontal table. An inextensible string in line with and fastened to the ruler passes perpendicularly over the table edge to support a mass m_2. A uniform rough sphere of mass M is set rolling along the ruler with velocity v away from the edge, and the ruler is released. Show that m_2 descends with constant acceleration and that the centre of the sphere comes to rest when m_2 has fallen a distance

$$7v^2(2M+7m_1+7m_2)/8m_2g.$$

3. Two particles of masses m and M are connected by a string passing through a small hole at the vertex of a cone having its axis vertical and vertex uppermost. If M hangs vertically, find the condition that m may describe a circle of radius C on the cone, and

show that if the particle is slightly disturbed it will oscillate about the circular path with period

$$2\pi\sqrt{\left[\frac{c(m+M)}{3g(M-m\cos\alpha)\sin\alpha}\right]}$$

where α is the half-angle of the cone.

4. Two equal rods AB and BC of length $2a$ and mass m are smoothly jointed together at B and lie at rest in a straight line on a smooth horizontal table. The end A of the system is suddenly seized and constrained to move with constant velocity V in a straight line through A perpendicular to the initial direction of ABC. Show that subsequently

$$4\dot{\theta}^2+\dot{\phi}^2+3\dot{\theta}\dot{\phi}\cos(\theta-\phi) = \frac{9}{7}\left(\frac{V}{a}\right)^2,$$

where θ is the angle between AB and its initial position and ϕ is similarly defined for BC. It may be assumed that A and C never coincide.

5. A uniform square lamina of side $2a$ and mass m is mounted on a corner O, which is fixed, in such a way that the lamina is always in some vertical plane but completely free otherwise. If θ is the angle between the diagonal through O and the vertical, and ϕ is the angle between the plane of the lamina and some fixed vertical plane through O, show that the kinetic energy T may be expressed in the form.

$$T = \tfrac{1}{6}ma^2[8\dot{\theta}^2+\dot{\phi}^2(1+6\sin^2\theta)].$$

Initially the lamina is set spinning with angular velocity Ω about the diagonal through O, held vertical momentarily. Show that $\dot{\phi}$ varies between Ω and $(1/7)\Omega$ and find the value of θ when $\dot{\phi}$ attains its minimum value.

6. A ring of wire of radius c rests horizontally on the top of a smooth fixed sphere of radius a, and is set rotating round the vertical diameter of the sphere. Show that this motion is unstable if the angular velocity is less than

$$[2g(2a^2-c^2)(a^2-c^2)^{\frac{1}{2}}c^{-4}]^{\frac{1}{2}}.$$

7. Show that the kinetic energy T of a symmetrical top can be expressed in the form

$$T = T_G + \tfrac{1}{2}I(\dot{\theta}^2 + \dot{\phi}^2 \sin^2 \theta) + \tfrac{1}{2}I_3(\dot{\phi} \cos \theta + \dot{\psi})^2,$$

where T_G is the kinetic energy of translation of the top, associated with the motion of its centre of mass G.

A symmetrical top is set spinning on a smooth horizontal table with angular velocity ω about its axis of symmetry which is vertical. The point of contact O of the top is on the axis of symmetry and a distance h below G. Show that the motion is stable if ω exceeds a certain critical value, and compare the stability with that of similar motion on a perfectly rough table.

8. Three equal uniform rods, AB, BC and CD, each of mass m, are freely jointed at B and C and rest in the form of three sides of a square $ABCD$ on a smooth horizontal table. If the ends A and D are suddenly made to move in the direction AD with velocities U and V respectively, determine the initial velocity of the rod BC and show that the kinetic energy of the system immediately after motion begins is

$$\frac{m}{120}\{19(U^2+V^2)-2UV\}.$$

9. Three equal uniform rods AB, BC, CD freely jointed at B and C are placed so as to form three sides of a square $ABCD$. If A is suddenly jerked parallel to the direction CB, show that the angular velocity of AB is initially seven times that of CD.

10. Two uniform rods AB, BC of equal length $2a$ and equal mass are smoothly jointed at B and rest on a smooth horizontal table with their directions at right angles to one another. A horizontal impulse is applied to the midpoint of AB and the rods start moving so that their angular velocities are equal and opposite. Find the direction of the impulse and prove that the initial speeds of A and C are in the ratio $\sqrt{13}:1$.

2 SMALL OSCILLATIONS

Theory of normal modes of small oscillations

Before giving the solution to any specific problem on normal modes, it will be useful to review the general theory of small oscillations. We tend to shy away from making use of powerful mathematical methods, but in the solution of a typical normal mode problem, unless one is dealing with a system possessing only two degrees of freedom, application of 'first principles' can be very tedious. Here the natural mathematical tool is the matrix theory of linear algebraic equations.

We consider a system of n degrees of freedom, and time-independent constraints. T, the kinetic energy of the system, is a quadratic form in the generalised velocities. We shall denote by $\dot{q}$ the column matrix of n elements $(\dot{q}_1, \dot{q}_2, \dots \dot{q}_n)$. Then T can be expressed in general as

$$T = \tfrac{1}{2}\dot{q}'A\dot{q}$$

where $\dot{q}'$ is the transpose of $\dot{q}$ and so is a row matrix. $A = (a_{ij})$, with $a_{ij} = a_{ij}(q_1, q_2, \dots q_n)$.

We assume that the forces acting on the system are derivable from a potential function $V(q_1, \dots q_n)$, which we shall write as $V(q)$. The generalised force Q will have n components $-(\partial V/\partial q_i)$, $i = 1, \dots n$.

Equilibrium. Suppose the system is in static equilibrium at $q = c$, $\dot{q} = 0$. (Once and for all, $q = c$, and any such equation, means a set of n equations, $q_i = c_i$). This is equivalent to the condition:

$$\text{at } q = c, \quad Q_i = -\frac{\partial V}{\partial q_i} = 0, \quad \text{all } i.$$

Now introduce a small displacement $\eta = q - c$. For small η,

$$V(q) = V(c+\eta) = V(c) + \sum_{i=1}^{n} \eta_i \left.\frac{\partial V}{\partial q_i}\right|_{q=c} + \tfrac{1}{2}\sum_{i,j=1}^{n} \eta_i\eta_j \left.\frac{\partial^2 V}{\partial q_i \partial q_j}\right|_{q=c} + \text{higher order terms.}$$

The terms linear in η vanish by definition of the equilibrium, while $V(c)$ may be omitted by appropriately defining the zero point of potential energy. Thus, close to equilibrium, we may write

$$V(q) \cong \tfrac{1}{2}\eta' B\eta$$

where

$$B = (b_{ij}), \quad b_{ij} = \left.\frac{\partial^2 V}{\partial q_i\, \partial q_j}\right|_{q=c}.$$

In the neighbourhood of $q = c$, A may also be taken to be constant. Further, A and B may be assumed symmetric, for their off-diagonal elements appear only added together, and may therefore be taken equal without changing the value of the quadratic forms T and V.

Since

$$\dot{c} = 0, \quad T = \tfrac{1}{2}\dot{\eta}' A\dot{\eta}.$$

For small departures from equilibrium, the Lagrangian L is given by

$$L = T - V = \tfrac{1}{2}(\dot{\eta}' A\dot{\eta} - \eta' B\eta),$$

and the equations of motion are

$$\frac{d}{dt}\left(\frac{\partial L}{\partial \dot{\eta}_i}\right) = \frac{\partial L}{\partial \eta_i}, \quad i = 1, \ldots n.$$

These can be written compactly as

$$A\,\ddot{} = -B\eta, \tag{1}$$

and form a set of n coupled second order linear differential equations with constant coefficients. To solve them, we suppose

$$\eta = C\,\mathrm{e}^{ipt}.$$

C, at this stage, is a column matrix of constant complex elements, and we are to understand the physical η to be the real part of $C\,\mathrm{e}^{ipt}$. Substituting into (*1*), we have

$$\ddot{\eta} = -p^2\eta.$$

Hence,

$$p^2 AC = BC. \tag{2}$$

Equation (*2*) constitutes a set of linear homogeneous simultaneous equations in C. For a non-trivial solution, we require

$$det(p^2 A - B) = 0.$$

This is known as the *secular determinant* and will generally be a polynomial of degree n in p^2. Since this is not meant to be an exhaustive dissertation we shall make, here and there, appropriate assumptions. Thus we shall assume that the roots are all distinct, and denote them by $p_1^2, \dots p_n^2$, no two roots equal; $p_1, \dots p_n$ are known as the *natural frequencies* of the system. The roots p_k^2 are real because A and B are real symmetric matrices. We assume further that all $p_k^2 \geqq 0$, for a negative p_k^2 implies that solutions are in general unbounded. We state in such a case that the system is *linearly unstable.*

To any p_k^2 there will correspond a column matrix C^k, defined up to an arbitrary complex constant since only *ratios* of elements C_i^k of C^k are defined. C^k satisfies the equation

$$BC^k = p_k^2 AC^k, \tag{3}$$

Then the general solution is

$$\eta = \sum_{k=1}^{n} C^k \, e^{ip_k t}, \tag{4}$$

which is a linear superposition of solutions, since equations are linear. Note that there are n arbitrary complex constants in equation (*4*), that is, $2n$ arbitrary real constants, as required. These may be determined by initial conditions of the motion.

Normal coordinates. Suppose we had chosen a different set of generalised coordinates $\xi_1, \dots \xi_n$ such that one set is a linear combination of the other,

$$\eta = G\xi.$$

In general we would have to carry out a similar calculation to the one we have outlined above, except that A and B would be different for the ξ coordinates. However, the natural frequencies would be unchanged, since the physical system is unchanged.

Suppose that, in terms of the new coordinates, the kinetic and potential energies were to take on a much simpler form, with no coupling terms between different coordinates, namely,

$$T = \tfrac{1}{2}\sum_{i=1}^{n} e_i \dot{\xi}_i^2 = \tfrac{1}{2}\dot{\xi}' E \dot{\xi},$$

$$V = \tfrac{1}{2}\sum_{i=1}^{n} f_i \xi_i^2 = \tfrac{1}{2}\xi' F \xi.$$

E and F are thus *diagonal* matrices. Then the equations of motion would become very simple.

$$e_k \ddot{\xi}_k = -f_k \xi_k, \quad \text{all } k,$$

with general solution

$$\xi_k = C_k \, e^{ip_k t},$$

where $p_k^2 = f_k/e_k$. The secular determinant, not needed here, would now take the form

$$\prod_{k=1}^{n} (f_k - p^2 e_k) = 0.$$

Physically, each generalised coordinate ξ_k oscillates only with the single frequency p_k. This motion is independent of that associated with any other ξ_i and the arbitrary constant C_k is determined entirely by the initial conditions on ξ_k, $\dot{\xi}_k$ and nothing else. The choice of ξ has thus decoupled the motion into n independently oscillating modes, and the ξ_k thus chosen are known as the *normal coordinates*.

So far, we have only *supposed* the existence of such normal coordinates. We now show plausibly how to construct them from any arbitrarily chosen set of generalised coordinates.

Let

$$\eta_i = \sum_{k=1}^{n} C_i^k \xi_k. \tag{5}$$

Note the analogy between equations (*4*) and (*5*). We have replaced exp $ip_k t$ in (*4*) by ξ_k, so we might expect that ξ_k will have the desired property of oscillating purely with frequency p_k. Let us now write (*5*) as

$$\eta = G\xi,$$

where $G = (g_{ik}) = (C_i^k)$. G can be regarded as a partitioned matrix with elements the column matrices C^k. We now want to express T and V in terms of $\dot{\xi}$ and ξ.

$$T = \tfrac{1}{2}\dot{\eta}'A\dot{\eta} = \tfrac{1}{2}\dot{\xi}'G'AG\dot{\xi} = \tfrac{1}{2}\dot{\xi}'E\dot{\xi},$$
$$V = \tfrac{1}{2}\eta'B\eta = \tfrac{1}{2}\xi'G'BG\xi = \tfrac{1}{2}\xi'F\xi.$$

We show that E and F are diagonal. The n equations of form (3), namely,

$$BC^k = p_k^2 AC^k,$$

can be put into the form of a single equation

$$BG = AG\Lambda \tag{6}$$

where Λ is diagonal, with p_k^2 in the kth diagonal element, zero elsewhere. Pre-multiplying (6) by G' gives

$$G'BG = G'AG\Lambda,$$

i.e.

$$F = E\Lambda.$$

But $E' = (G'AG)' = G'A'G = G'AG = E$, and similarly $F' = F$. Hence,

$$(E\Lambda)' = E\Lambda,$$

i.e.

$$\Lambda E = E\Lambda.$$

Since Λ is diagonal, with distinct elements, so must E be diagonal, since

$$(\Lambda E)_{ij} = (E\Lambda)_{ij} \quad i \neq j,$$

i.e.

$$p_i^2\, e_{ij} = e_{ij} p_j^2 \Rightarrow e_{ij} = 0,$$

since

$$p_i^2 \neq p_j^2,$$

unless

$$i = j.$$

Since E is diagonal, so is F.

If n is small, say $n = 3$, we get three equations

$$\eta_1 = C_1^1\xi_1 + C_1^2\xi_2 + C_1^3\xi_3,$$
$$\eta_2 = C_2^1\xi_1 + C_2^2\xi_2 + C_2^3\xi_3,$$
$$\eta_3 = C_3^1\xi_1 + C_3^2\xi_2 + C_3^3\xi_3,$$

and it is simple to solve these for ξ in terms of η.

We may be faced with a computer calculation in which the system in question has perhaps thirty degrees of freedom. It is then much better to make use of certain properties of the C^k. Thus, we wish to invert the equation $\eta = G\xi$, to get ξ. On pre-multiplying by $G'A$, we obtain

$$G'A\eta = G'AG\xi.$$

The right hand term is $E\xi$, and E is diagonal with elements e_i. Thus

$$E\xi = G'A\eta,$$

i.e.

$$\xi = E^{-1}G'A\eta.$$

and E^{-1} is also diagonal with elements e_i^{-1}. It is then straightforward to obtain ξ by a method which involves only matrix multiplication. Such operations are very convenient for high speed digital com puters.

2.1. *A rod of length l and negligible mass is connected at one end to the centre of a uniform circular disc of mass m and radius a. The disc and the rod are coplanar and the connection between them has torsional rigidity k (i.e. a torque $k\theta$ on the disc about the normal through its centre produces an angle of twist θ). The system is suspended from the free end of the rod and performs small oscillations in a vertical plane coinciding with the plane of the disc. Show that the natural frequencies of the system are given by the equation*

$$p^4 - [\Omega^2 + (1 + a^2/2l^2)\omega^2]p^2 + \omega^2\Omega^2 = 0,$$

where $\Omega^2 = g/l$ and $\omega^2 = 2k/ma^2$.

Show graphically how these vary as k varies from 0 to ∞ and describe the physical meaning of the results.

We choose generalised coordinates θ and ϕ as in Fig. 10.

The kinetic energy is given by

$$T = \tfrac{1}{2}ml^2\dot{\phi}^2 + \tfrac{1}{4}ma^2(\dot{\theta} + \dot{\phi})^2.$$

On the right, the first term describes the motion of the mass centre O of the disc, the second refers to rotation about O.

$$\begin{aligned} V &= \text{potential energy} \\ &= \tfrac{1}{2}mgl\phi^2 + \tfrac{1}{2}k\theta^2. \end{aligned}$$

On the right, we have approximated $1-\cos\phi$ by $\frac{1}{2}\phi^2$. Introducing $\Omega^2 = g/l$ and $\omega^2 = 2k/ma^2$, we obtain

$$V = \tfrac{1}{2}m\Omega^2 l^2\phi^2 + \tfrac{1}{4}m\omega^2 a^2\theta^2.$$

To obtain the equation for the natural frequencies we write

$$T = X'AX \quad \text{and} \quad V = Y'BY$$

where X is a column matrix with elements $\dot{\theta}$, $\dot{\phi}$ and Y a column matrix with elements θ, ϕ. A and B are the appropriate symmetrical

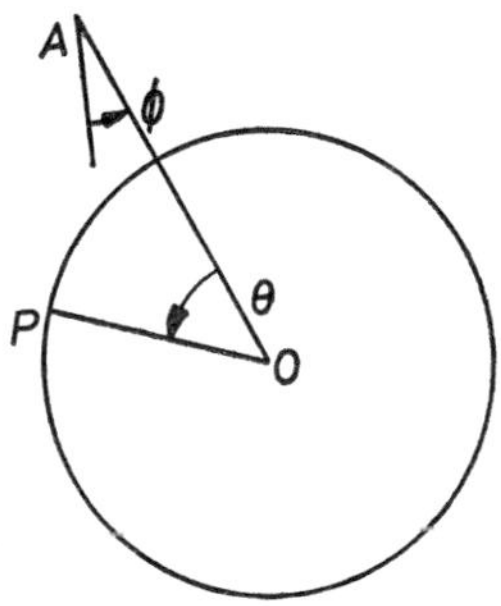

FIG. 10

2×2 matrices. The secular equation is then $det(p^2A-B) = 0$, namely

$$\tfrac{1}{4}a^2\begin{vmatrix} p^2-\omega^2 & p^2 \\ \dfrac{a^2}{2l^2}p^2 & \left(1+\dfrac{a^2}{2l^2}\right)p^2-\Omega^2 \end{vmatrix} = 0.$$

This simplifies to the required equation

$$p^4-[\Omega^2+(1+a^2/2l^2)\omega^2]p^2+\omega^2\Omega^2 = 0.$$

To discuss the properties of the natural frequencies as k is varied, we solve the equation for p^2

$$p^2 = \tfrac{1}{2}\{\Omega^2+(1+a^2/2l^2)\omega^2 \pm [\Omega^4+2(a^2/2l^2-1)\Omega^2\omega^2 + (1+a^2/2l^2)^2\omega^4]^{\frac{1}{2}}\}.$$

We are interested in the two limiting cases $k \to 0$ and $k \to \infty$, i.e. $\omega \ll \Omega$ and $\omega \gg \Omega$.

Case 1. $\omega \ll \Omega$.

$$p^2 \cong \tfrac{1}{2}\{\Omega^2+(1+a^2/2l^2)\omega^2 \pm \Omega^2[1+(a^2/2l^2-1)(\omega^2/\Omega^2)]\},$$

where we have expanded the square root, taking account of the fact that $\omega/\Omega \ll 1$. The two roots p_1^2 and p_2^2 are now seen to be given by

$$p_1^2 \cong \Omega^2, \quad p_2^2 \cong \omega^2.$$

Notice that we kept all terms of order ω^2/Ω^2 in obtaining an approximate solution, but now, in p_1^2 we have neglected any small corrections. They are required for p_2^2, which vanishes in lowest order.

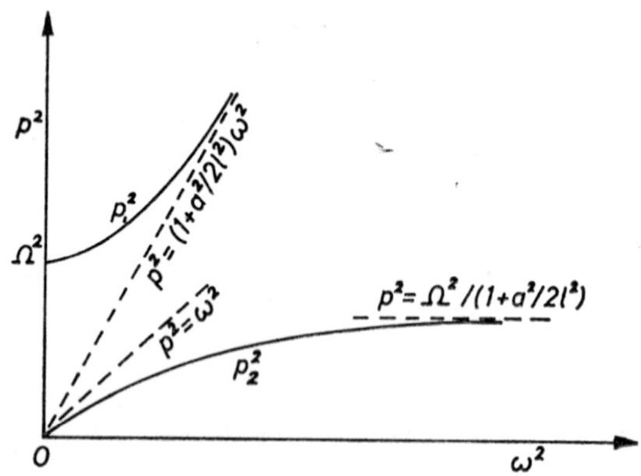

FIG. 11

Case 2. $\omega \gg \Omega$.

The reader may verify for himself that, to first order in Ω^2/ω^2, approximate solutions are

$$p_1^2 \cong (1+a^2/2l^2)\omega^2, \quad p_2^2 \cong \Omega^2/(1+a^2/2l^2).$$

The various solutions are shown in Fig. 11.

Note that the larger frequency p_1 has a minimum value equal to Ω, increasing indefinitely towards the asymptotic value $\omega(1+a^2/2l^2)^{\frac{1}{2}}$. The smaller frequency p_2 equals ω for small ω and increases to a limiting value $\Omega(1+a^2/2l^2)^{-\frac{1}{2}}$. It will be seen that there is a band of frequencies between $\Omega(1+a^2/2l^2)^{-\frac{1}{2}}$ and Ω which is unobtainable. The system could act as a filter, blocking out frequencies in this band.

When $k \to 0$, there is no coupling, so the disc rotates freely. The root $p_1 = \Omega$ corresponds to a simple pendulum of length l; $p_2 = \omega \to 0$ corresponds to a very slow oscillation independent of the motion of the pendulum.

When $k \to \infty$, the coupling is rigid and the motion is rigid body motion of the whole system. Now $p_1 \to \infty$ and becomes physically unobservable due to any frictional effects which will tend to damp out this mode; $p_2 \to \Omega/(1+a^2/2l^2)^{\frac{1}{2}}$, which is the oscillation frequency of the equivalent simple pendulum. This last statement should be verified.

It is important, for various reasons, that one should be able to relate limiting cases to simple physical situations. In the problem above, these limiting cases can be studied by elementary methods. This is a very powerful technique often used in theoretical research where, after perhaps much complicated analysis one considers various limiting cases which can be checked from first principles. Secondly, it is by studying such problems that one obtains more insight into the completeness of physics, where every piece of information about a particular system must fit logically into the whole. And finally, it is from this kind of work that unexpected results may appear. In this problem we find (unexpectedly?) a simple model for a frequency band filter.

2.2. *A light string OAB is tied to a fixed point at O, and carries a mass 2m at A and a mass m at B. The lengths OA, AB are $\frac{1}{2}l$, $\frac{3}{4}l$ respectively. The string is free to move in a vertical plane, and the system oscillates about the position of equilibrium. The inclinations of OA, AB to the vertical are denoted by θ, ϕ respectively. Find the normal coordinates.*

The system is held with the string straight and inclined at a small angle α to the vertical, and is let go from rest in this position at the instant $t = 0$. Show that at any subsequent time

$$\theta = \tfrac{1}{3}\alpha(2\cos nt + \cos 2nt),$$
$$\phi = \tfrac{1}{3}\alpha(4\cos nt - \cos 2nt),$$

where $n = (g/l)^{\frac{1}{2}}$.

For small angular displacements θ, ϕ,

$$V_A \cong \tfrac{1}{2}l\dot{\theta}, \quad V_B \cong \tfrac{1}{2}l\dot{\theta} + \tfrac{3}{4}l\dot{\phi}.$$

Hence,

$$T = \text{kinetic energy for small oscillations,}$$
$$= \tfrac{1}{2}\,.\,2m\,.\,(\tfrac{1}{2}l\dot{\theta})^2+\tfrac{1}{2}m(\tfrac{1}{2}l\dot{\theta}+\tfrac{3}{4}l\dot{\phi})^2,$$
$$= \tfrac{3}{32}ml^2(4\dot{\theta}^2+4\dot{\theta}\dot{\phi}+3\dot{\phi}^2).$$

$$V = \text{potential energy,}$$
$$= 2mg\,.\,\tfrac{1}{2}l(1-\cos\theta)+mg[\tfrac{3}{4}l(1-\cos\phi)+\tfrac{1}{2}l(1-\cos\theta)].$$

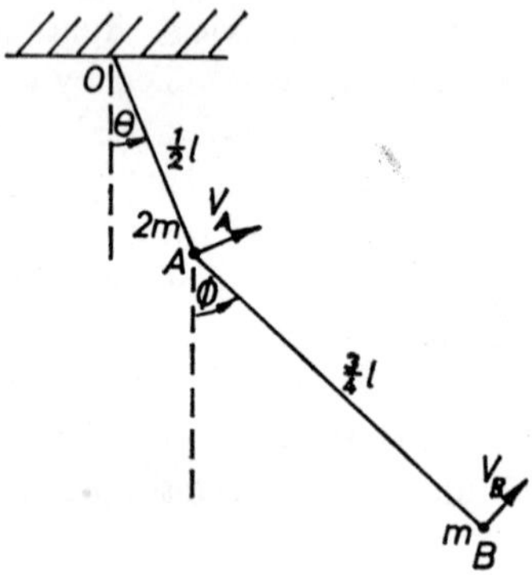

FIG. 12

Using the approximation $1-\cos\theta \cong \theta^2/2$, we obtain

$$V = \tfrac{3}{8}mgl(2\theta^2+\phi^2).$$

Let $n^2 = g/l$. Then the secular equation is

$$\begin{vmatrix} 4p^2-8n^2 & 2p^2 \\ 2p^2 & 3p^2-4n^2 \end{vmatrix} = 0,$$

i.e.

$$p^4-5p^2n^2+4n^4 = 0$$

with solution

$$p_1^2 = n^2, \quad p_2^2 = 4n^2.$$

Determination of C^k.

For C^1, either row gives $C_2^1 = 2C_1^1$, so we may take $C^1 = \begin{bmatrix}1\\2\end{bmatrix}$. Similarly,

$$C^2 = \begin{bmatrix}1\\-1\end{bmatrix}.$$

Normal coordinates ξ_1 and ξ_2 are given by

$$\left.\begin{aligned}\theta &= C_1^1\xi_1+C_1^2\xi_2 = \xi_1+\xi_2\\ \phi &= C_2^1\xi_1+C_2^2\xi_2 = 2\xi_1-\xi_2\end{aligned}\right\} \qquad (1)$$

Hence

$$\xi_1 = \tfrac{1}{3}(\theta+\phi), \quad \xi_2 = \tfrac{1}{3}(2\theta-\phi). \qquad (2)$$

Initially,

$$\theta = \phi = \alpha \quad \text{and} \quad \dot{\theta} = \dot{\phi} = 0.$$

Hence, using (*2*),

$$\xi_1 = \tfrac{2}{3}\alpha, \quad \xi_2 = \tfrac{1}{3}\alpha, \quad \dot{\xi}_1 = \dot{\xi}_2 = 0.$$

Thus at any later time t

$$\xi_1 = \tfrac{2}{3}\alpha \cos nt, \quad \xi_2 = \tfrac{1}{3}\alpha \cos 2nt,$$

and so, using (*1*),

$$\theta = \tfrac{1}{3}\alpha(2 \cos nt+\cos 2nt),$$

$$\phi = \tfrac{1}{3}\alpha(4 \cos nt-\cos 2nt).$$

We shall now obtain these results by more elementary methods. The analysis determining normal modes is very elegant but some students see it more as a recipe for obtaining solutions, while the physical meaning of normal coordinates is partially obscured. (The authors' opinion would contradict this claim!) We shall assume that the method of generalised coordinates has been used to obtain the expression given above for T and V, for small angular displacements θ and ϕ, namely,

$$T = \tfrac{3}{32}ml^2(4\dot{\theta}^2+4\dot{\theta}\dot{\phi}+3\dot{\phi}^2),$$

$$V = \tfrac{3}{8}mgl(2\theta^2+\phi^2).$$

The equations of motion are

$$\frac{d}{dt}\left(\frac{\partial T}{\partial \dot{\theta}}\right) = \frac{\partial T}{\partial \theta} - \frac{\partial V}{\partial \theta} = -\frac{\partial V}{\partial \theta},$$

and similarly for ϕ, i.e.

$$\tfrac{3}{32}ml^2(8\ddot{\theta}+4\ddot{\phi}) = -\tfrac{3}{2}mgl\theta,$$

$$\tfrac{3}{32}ml^2(4\ddot{\theta}+6\ddot{\phi}) = -\tfrac{3}{4}mgl\phi.$$

Let $n^2 = g/l$. The equations simplify to

$$2\ddot{\theta}+\ddot{\phi}+4n^2\theta = 0,$$

$$2\ddot{\theta}+3\ddot{\phi}+4n^2\phi = 0.$$

Assume trial solutions

$$\theta = C\cos(pt+\alpha), \quad \phi = D\cos(pt+\alpha).$$

The object is to find values of p which do not lead to the trivial solution $C = D = 0$. Substitution into the equations of motion gives

$$2Cp^2+Dp^2-4Cn^2 = 0,$$

$$2Cp^2+3Dp^2-4Dn^2 = 0.$$

Hence

$$D = C(4n^2-2p^2)/p^2, \tag{3}$$

and

$$2Cp^2+C(4n^2-2p^2)(3p^2-4n^2)/p^2 = 0.$$

If C is to be non-zero, the equation above requires

$$p^4-5p^2n^2+4n^4 = 0,$$

which is of course just the secular equation with roots $p_1 = n$, $p_2 = 2n$. Equation (3) is now the all-important equation to be used for determining the normal coordinates; for any given value of p, the ratio D/C is defined. In particular, we find, for

$$p_1 = n, \quad D = 2C$$

and

$$p_2 = 2n, \quad D = -C,$$

with corresponding solutions, respectively,

$$\theta = C\cos nt, \quad \phi = 2C\cos nt$$

and

$$\theta = C\cos 2nt, \quad \phi = -C\cos 2nt.$$

Thus the relation between C and D is precisely what is contained in the C^k of general theory. For the normal mode with frequency n, $\theta = \frac{1}{2}\phi$. For the mode with frequency $2n$, $\theta = -\phi$. Any general motion will be a linear superposition of these two pure oscillations, namely,

$$\theta = C_1\cos(nt+\alpha_1)+C_2\cos(2nt+\alpha_2),$$

$$\phi = 2C_1\cos(nt+\alpha_1)-C_2\cos(2nt+\alpha_2).$$

Note carefully that C_1 and C_2 are arbitrary constants (and α_1, α_2 arbitrary phases), but that once these are given for θ the expression for ϕ is completely defined, each part having the appropriate amplitude and being either in (or out of) phase with the corresponding term in θ.

At this point, the initial conditions define the arbitrary amplitudes and phases, giving the same result as before.

It should be stressed that this part has been included to explain in detail the various points which occurred in the earlier more elegant method. However, one of the lessons to learn is that 'elementary' methods tend to be tedious because of their inevitably longer exposition. We need to practice the apparently more sophisticated mathematics until it itself becomes more 'elementary'. It is not introduced for its own sake, but because of its power in solving problems.

2.3. *Two particles, each of mass 3m, lie on either side of a particle of mass 2m, to which they are connected by means of identical springs of stiffness k. The particles lie in a straight line on a smooth horizontal table and are constrained to move only in that line. Show that the natural frequencies of small oscillations of the system are* 0, ω, 2ω, *where* $\omega^2 = k/3m$.

Find the normal coordinates, discuss their physical meaning, and verify that in any general motion involving small oscillations the total linear momentum of the system is conserved.

In Fig. 13, η_1, η_2 and η_3 are the displacements of the masses, as shown, from their equilibrium positions. When we speak of *small* oscillations we shall of course be interested only in the smallness of $|\eta_1 - \eta_2|$ and $|\eta_2 - \eta_3|$, not in the absolute values of η_1, η_2 and η_3.

In this problem, we have first to obtain the natural frequencies. This involves the determination of the kinetic and potential energies

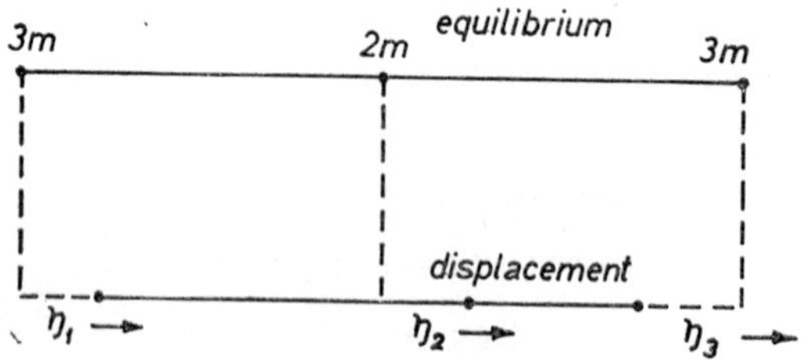

FIG. 13

in terms of generalised coordinates and velocities. For generalised coordinates we shall of course choose η_1, η_2 and η_3.

$$T = \text{kinetic energy},$$
$$= \tfrac{1}{2}[3m\dot{\eta}_1^2 + 2m\,\dot{}_2^2 + 3m\dot{\eta}_3^2].$$

Obtaining the potential energy V can sometimes prove difficult. From the diagram it is easy to obtain the forces acting on each particle. Thus, on the left hand particle of mass $3m$, the force is $k(\eta_2 - \eta_1)$, towards the right if $\eta_2 > \eta_1$. For the central particle of mass $2m$, the total force is $k(\eta_3 - \eta_2) - k(\eta_2 - \eta_1)$, and so on. The property of $V(\eta_1, \eta_2, \eta_3)$ is that the force on the particle with displacement coordinate η_i is $-(\partial V/\partial \eta_i)$. One should see from these considerations that

$$V = \tfrac{1}{2}k[(\eta_2 - \eta_1)^2 + (\eta_3 - \eta_2)^2]$$

satisfies the required properties. It is possible to obtain V by a different method, by noting that the potential energy of a spring of stiffness k stretched by an amount x is just $\frac{1}{2}kx^2$. Each term in V corresponds to the potential energy of each spring.

Letting $\omega^2 = k/3m$, we have

$$T = \tfrac{1}{2}m(3\dot{\eta}_1^2 + 2\dot{\eta}_2^2 + 3\dot{\eta}_3^2),$$
$$V = \tfrac{3}{2}m\omega^2(\eta_1^2 + 2\eta_2^2 + \eta_3^2 - 2\eta_1\eta_2 - 2\eta_2\eta_3).$$

The secular equation is therefore

$$\begin{vmatrix} p^2-\omega^2 & \omega^2 & 0 \\ \omega^2 & \frac{2}{3}p^2-2\omega^2 & \omega^2 \\ 0 & \omega^2 & p^2-\omega^2 \end{vmatrix} = 0,$$

i.e.

$$(p^2-\omega^2)[(p^2-2\omega^2)(p^2-\omega^2)-2\omega^4] = 0,$$

i.e.

$$p^2(p^2-\omega^2)(p^2-4\omega^2) = 0.$$

Thus the roots of the secular equation are $p = 0, \omega, 2\omega$. To obtain the normal coordinates, we need the latent vectors C^k corresponding to the roots $0, \omega, 2\omega$. The property of C^k is

$$(p_k^2A-B)C^k = 0.$$

For $C^1, p_1 = 0$,

$$\begin{bmatrix} -\omega^2 & \omega^2 & 0 \\ \omega^2 & -2\omega^2 & \omega^2 \\ 0 & \omega^2 & -\omega^2 \end{bmatrix} \begin{bmatrix} C_1^1 \\ C_2^1 \\ C_3^1 \end{bmatrix} = 0,$$

and since the determinant $det(p_1^2A-B) = 0$, we are guaranteed a non-trivial solution.

From the first row-multiplication, we have

$$C_1^1 = C_2^1,$$

and likewise the third row-multiplication yields

$$C_2^1 = C_3^1,$$

so a convenient

$$C^1 = \begin{bmatrix} 1 \\ 1 \\ 1 \end{bmatrix}.$$

For $C^2, p_2 = \omega$,

$$\begin{bmatrix} 0 & \omega^2 & 0 \\ \omega^2 & -\frac{4}{3}\omega^2 & \omega^2 \\ 0 & \omega^2 & 0 \end{bmatrix} \begin{bmatrix} C_1^2 \\ C_2^2 \\ C_3^2 \end{bmatrix},$$

and

$$C^2 = \begin{bmatrix} 1 \\ 0 \\ -1 \end{bmatrix}.$$

Verify that for C^3, corresponding to $p_3 = 2\omega$, we find

$$C^3 = \begin{bmatrix} 1 \\ -3 \\ 1 \end{bmatrix}.$$

Having obtained the matrix $C = [C^1 C^2 C^3]$ we are now in a position to determine the normal coordinates ξ_1, ξ_2 and ξ_3, which satisfy the property

$$\eta = C\xi.$$

Thus

$$\begin{bmatrix} \eta_1 \\ \eta_2 \\ \eta_3 \end{bmatrix} = \begin{bmatrix} 1 & 1 & 1 \\ 1 & 0 & -3 \\ 1 & -1 & 1 \end{bmatrix} \begin{bmatrix} \xi_1 \\ \xi_2 \\ \xi_3 \end{bmatrix},$$

i.e.

$$\eta_1 = \xi_1 + \xi_2 + \xi_3, \tag{1}$$

$$\eta_2 = \xi_1 - 3\xi_3, \tag{2}$$

$$\eta_3 = \xi_1 - \xi_2 + \xi_3. \tag{3}$$

It remains only to invert these equations to find the ξ_i in terms of the η_k. For a system of only three degrees of freedom, one need not apply the general properties of the matrix C (though it would be instructive to see for oneself if one could do so). Instead, a direct algebraic inversion is all that is required here. The result is

$$\xi_1 = \tfrac{1}{8}(3\eta_1 + 2\eta_2 + 3\eta_3), \tag{4}$$

$$\xi_2 = \tfrac{1}{2}(\eta_1 - \eta_3), \tag{5}$$

$$\xi_3 = \tfrac{1}{8}(\eta_1 - 2\eta_2 + \eta_3). \tag{6}$$

The easiest way to find the physical nature of the normal coordinates is to consider the motion of the system when, in turn, only one of the normal coordinates is non-zero. Remember that the essential property of the normal coordinates is that they are uncoupled to each other, so that each oscillates independently at its own particular frequency p_k. We can therefore consider the special case where only one ξ_k oscillates. The physical nature is best seen using equations (*1*), (*2*) and (*3*).

$$\xi_1 = x, \quad \xi_2 = \xi_3 = 0 \rightarrow \eta_1 = \eta_2 = \eta_3 = x.$$

Thus ξ_1 corresponds to the case where particles are all equally displaced.

$$\xi_2 = x, \quad \xi_1 = \xi_3 = 0 \rightarrow \eta_1 = x, \quad \eta_2 = 0, \quad \eta_3 = -x.$$

Now the central particle is stationary, and the outer particles are displaced in opposite directions by equal amounts.

$$\xi_3 = x, \quad \xi_1 = \xi_2 = 0 \rightarrow \eta_1 = \eta_3 = x, \quad \eta_2 = -3x.$$

In this final mode, the two outer particles are equally displaced, while the central particle suffers a displacement three times that of the outer particles but in the opposite direction.

The total linear momentum P is given by

$$P = 3m\dot{\eta}_1 + 2m\dot{\eta}_2 + 3m\dot{\eta}_3.$$

From equation (*4*), $P = 8m\dot{\xi}_1$, and the equation for ξ_1 is $\ddot{\xi}_1 + p_1^2\xi_1 = 0$, i.e. $\ddot{\xi}_1 = 0$. Hence $\dot{\xi}_1$ is constant and P is conserved. Notice that the conservation of linear momentum is reflected in the existence of a zero root of the secular equation. The result can also be seen when we realise that the ξ_1 mode corresponds to a pure translation of the system, with no change in the potential energy since $\eta_1 - \eta_2$ and $\eta_2 - \eta_3$ are zero. Verify that in the other two modes only *internal* motion is involved, the centre of mass of the system remaining stationary.

EXERCISES

1. A uniform, thin-walled, hollow cylinder of radius a and mass M, is free to roll on a rough horizontal plane. A particle of mass m slides on the smooth inner surface of the cylinder. Show that for small motions

$$T = (M + \tfrac{1}{2}m)\dot{x}^2 + \tfrac{1}{2}ma^2\dot{\theta}^2 + ma\dot{x}\dot{\theta},$$

and

$$V = \tfrac{1}{2}mga\theta^2,$$

where x is the displacement of the cylinder and θ is the angular displacement of the particle from the vertical. Hence determine the period of oscillation.

2. A uniform light spring is hung from a fixed support, and similar masses m are fixed at the mid-point and lower end of it. Find an expression for the frequencies of normal modes of vertical oscillation in terms of m and the elastic constant of the spring. Describe the normal modes and find the ratios of the amplitudes of oscillation of the two masses.

3. The balance wheel of a fixed watch undergoes a rotary oscillation in its own plane and about its own centre, with simple harmonic motion of period $2\pi/p$. The consequent kinetic and potential energies are given by $\tfrac{1}{2}J\dot{\phi}^2$ and $\tfrac{1}{2}Jp^2\phi^2$ respectively, where J is the moment of inertia of the wheel about an axis through its centre normal to its plane, and ϕ is the angle of rotation.

The watch is now suspended so that it can perform small oscillations in the vertical plane containing the balance wheel. Let I be the moment of inertia of the whole watch (including the balance wheel) about an axis through the point of suspension normal to the plane of oscillation and let θ be the angular displacement of the watch from its equilibrium position. Show that the Langrangian for the whole watch is now given by

$$L = \tfrac{1}{2}[(I-J)\dot{\theta}^2+J(\dot{\theta}+\dot{\phi})^2]-\tfrac{1}{2}[Iq^2\theta^2+Jp^2\phi^2],$$

where q is a constant. Find the period of normal oscillation to the first order in the small quantity J/I. Does the watch gain or lose when thus suspended?

4. A smooth circular wire of mass $8m$ and radius a swings in a vertical plane, being suspended by an inextensible string of length a attached to one point of it. A bead of mass m can slide on the wire. Show that the angle between the radius to the bead and the diameter through the point of suspension is a normal coordinate. Prove that the periods of the normal oscillations are

$$2\pi\sqrt{\frac{8a}{3g}},\quad 2\pi\sqrt{\frac{a}{3g}},\quad 2\pi\sqrt{\frac{8a}{9g}}$$

and find the other normal coordinates. Illustrate by sketches the nature of each normal mode.

5. Three particles, two of mass m, and the third of mass λm, are equally spaced in a smooth circular groove of radius a.

The particles are connected by three identical springs which fit into the groove. Discuss the nature of the normal modes of small oscillations of the system.

The system is initially at rest in equilibrium. The particle of mass λm is subjected to a blow which gives it a small instantaneous velocity V tangential to the groove. Calculate the maximum velocity attained by the other particles in the subsequent motion of the system.

6. A uniform square lamina, mass M, side $2a$, is supported at its corners by four equal springs such that in equilibrium the plane of the lamina is horizontal. The lamina is caused to undergo small oscillations about its position of equilibrium and the vertical displacements from the equilibrium position of the corners X_1, X_2, X_3, X_4 are x_1 x_2, x_3 and x_4 respectively. Show that the Langrangian can be written in the form

$$L = \frac{M}{2}\left(\frac{\dot{x}_1+\dot{x}_3}{2}\right)^2 + \frac{M}{48}[(\dot{x}_1-\dot{x}_3)^2+(\dot{x}_2-\dot{x}_4)^2] - \frac{k}{2}(x_1+x_3-2x_0)^2$$

$$-\frac{k}{4}(x_1-x_3)^2 - \frac{k}{4}(x_2-x_4)^2 - \frac{Mg}{2}(x_1+x_3),$$

where X_1X_3 and X_2X_4 are the diagonals of the square and $\frac{1}{2}k(x_i-x_0)^2$ is the potential energy of the spring at x_i. Hence show that there are only two characteristic frequencies of vibration $2\pi\sqrt{(x_0/g)}$, $2\pi\sqrt{(x_0/3g)}$ and account for the degeneracy.

3 VECTOR DYNAMICS OF RIGID BODIES

It is not possible as an introduction to this chapter to give a full account of the principles of vector dynamics; on the other hand a short summary does not serve much purpose. It is best to let the examples in the text speak for themselves. Suffice it to say that, as we have seen in Chapter 1, a rigid body moving without constraints possesses six degrees of freedom and thus requires six scalar equations of motion. In vector dynamics these usually take the form of the two *vector* equations

$$\frac{d\mathbf{P}}{dt} = \mathbf{F}, \quad \frac{d\mathbf{L}}{dt} = \mathbf{N},$$

where **P** and **L** are the total linear and angular momentum, the latter measured either with respect to a fixed origin 0 or the mass centre G. **P** on the other hand is measured from 0. (Slight variations from this scheme are possible, but will not be discussed here). **F** and **N** are the total external force and torque, the latter about the same origin as the corresponding angular momentum. For rotation about a fixed point, only the angular momentum equation is needed, **L** being taken about the point of rotation.

Often rolling constraints have to be applied, and the simplest general advice in deriving a rolling constraint is to consider the point of contact simultaneously as a point of the rigid body and also as a point of the surface on which it is rolling. Then if no slip occurs the velocity of this point is the same from both points of view. If the surface is stationary, the point of contact is instantaneously at rest; if the surface is for example a rotating plane, then the point of contact (instantaneously) rotates with appropriate angular velocity about the axis of rotation of the plane.

3.1. *A square lamina ABCD of mass M and side 2a is rotating about the side AB with angular velocity ω. Show that the angular momentum about the point D is given by*

$$\boldsymbol{L}_D = -\tfrac{2}{3}Ma^2\omega\boldsymbol{k} - Ma^2\omega\boldsymbol{i},$$

where **k**, ***i*** *are unit vectors in the appropriate sense along the sides AB, BC respectively.*

The axis AB is released and the point D is simultaneously seized. Calculate the magnitude and direction of the instantaneous angular velocity about D.

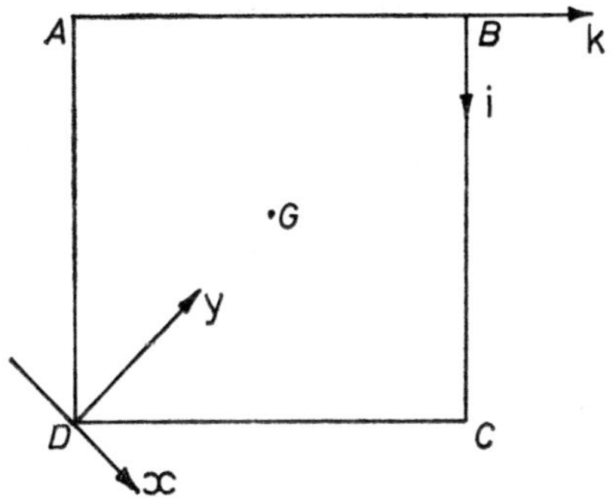

FIG. 14

This problem illustrates general principles involved in the discussion of impulsive motion. The sudden change in the motion of the lamina is produced by the action of an impulsive force acting at D (since it is the point seized). Since the impulse acts at D, there can be no torque about D. Thus the angular momentum of the lamina about D must remain unchanged. The first part of the problem involves the calculation of $\boldsymbol{L}_D$ just before D is fixed. You may imagine that the point D is that point fixed in space which just coincides with the moving point D of the lamina just as it is seized. In the second part we have to determine $\boldsymbol{L}_D$, assuming that the lamina is rotating about D with angular velocity $\boldsymbol{\omega}'$. By equating this second expression for $\boldsymbol{L}_D$ to the first, we shall be able to determine $\boldsymbol{\omega}'$.

We first determine $\boldsymbol{L}_G$, and then use the relation

$$\boldsymbol{L}_D = \boldsymbol{L}_G + \overrightarrow{DG} \times \boldsymbol{P},$$

where $\boldsymbol{P}$ is the total momentum of the lamina.

Since any axis in the plane of the lamina through G is a principal axis, with moment of inertia $\frac{1}{3}Ma^2$

$$\boldsymbol{L}_G = \tfrac{1}{3}Ma^2\omega\boldsymbol{k}.$$

Hence,

$$\begin{aligned}\boldsymbol{L}_D &= \tfrac{1}{3}Ma^2\omega\boldsymbol{k}+a(\boldsymbol{k}-\boldsymbol{i})\times Ma\omega\boldsymbol{j},\\ &= -Ma^2\omega(\tfrac{2}{3}\boldsymbol{k}+\boldsymbol{i}).\end{aligned}$$

The principal axes at D are along the diagonal DB and perpendicular to DB; $\boldsymbol{x}$ and $\boldsymbol{y}$ are unit vectors along these principal axes. We can therefore write

$$\boldsymbol{L}_D = -Ma^2\omega[\tfrac{2}{3}(\boldsymbol{x}+\boldsymbol{y})/\sqrt{2}+(\boldsymbol{x}-\boldsymbol{y})/\sqrt{2}].$$

After D is seized, let the angular velocity of the lamina be $\boldsymbol{\omega}' = \omega_1\boldsymbol{x}+\omega_2\boldsymbol{y}$. Hence,

$$\boldsymbol{L}_D = \tfrac{7}{3}Ma^2\omega_1\boldsymbol{x}+\tfrac{1}{3}Ma^2\omega_2\boldsymbol{y}.$$

Here we have used the property of principal axes, and the values of the principal moments of inertia about D (which should be checked by the reader). We may now apply the conservation of angular momentum about D, equating the coefficients of $\boldsymbol{x}$ and $\boldsymbol{y}$ to obtain ω_1 and ω_2.

Thus,

$$\begin{aligned}\tfrac{7}{3}Ma^2\omega_1 &= -5Ma^2\omega/3\sqrt{2},\\ \tfrac{1}{3}Ma^2\omega_2 &= Ma^2\omega/3\sqrt{2},\end{aligned}$$

i.e.

$$\omega_1 = -5\omega/7\sqrt{2}, \quad \omega_2 = \omega/\sqrt{2}.$$

Reverting to the original unit vectors $\boldsymbol{i}, \boldsymbol{k}$, we obtain

$$\begin{aligned}\boldsymbol{\omega}' &= \omega(-\tfrac{5}{7}\boldsymbol{x}+\boldsymbol{y})/\sqrt{2},\\ &= \omega(6\boldsymbol{i}-\boldsymbol{k})/7.\end{aligned}$$

This defines $\boldsymbol{\omega}'$ in magnitude and direction.

3.2. *Show that the centre of a sphere, rolling without slipping on an inclined plane, traces out a parabola.*

Problems of this type are invariably solved by setting down three equations, namely: an equation for linear momentum, one for angular momentum, and one which expresses the rolling condition.

Referring to the diagram, $\boldsymbol{n}$ is the unit normal to the inclined plane, O a fixed point on the plane, P the point of contact of sphere and plane, and C the centre of the sphere. Forces acting on the sphere

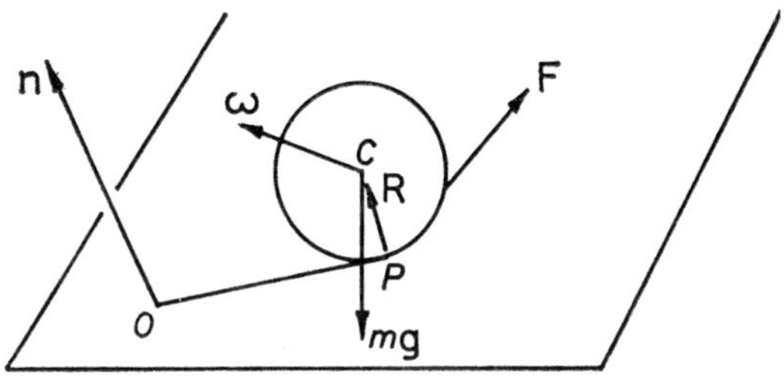

FIG. 15

are the gravitational force mg acting vertically downwards through C, and the reaction of the plane on the sphere; the reaction is expressed as a component $\boldsymbol{F}$ in the plane and a normal reaction $\boldsymbol{R} = R\boldsymbol{n}$. The angular velocity of the sphere is $\boldsymbol{\omega}$. Let $\boldsymbol{r}_P = \boldsymbol{r}$. Then $\boldsymbol{r}_C = \boldsymbol{r}+a\boldsymbol{n}$, $\dot{\boldsymbol{r}}_C = \dot{\boldsymbol{r}}$ and $\ddot{\boldsymbol{r}}_C = \ddot{\boldsymbol{r}}$.

Equation of linear momentum:

$$m\ddot{\boldsymbol{r}} = R\boldsymbol{n}+\boldsymbol{F}+m\boldsymbol{g} \tag{1}$$

Equation of angular momentum:

$$mk^2\dot{\boldsymbol{\omega}} = \overrightarrow{CP}\times\boldsymbol{F}$$

$$= -a\boldsymbol{n}\times\boldsymbol{F} \tag{2}$$

Rolling condition:

$$\dot{\boldsymbol{r}}+\boldsymbol{\omega}\times\overrightarrow{CP} = 0,$$

i.e.

$$\dot{\boldsymbol{r}}-a\boldsymbol{\omega}\times\boldsymbol{n} = 0. \tag{3}$$

The left hand side of equation (*3*) is the velocity of P considered as a point on the surface of the sphere. This velocity must be equated to zero since there is no slip between the sphere and the (fixed) plane.

If the sphere is to remain in contact with the plane, $\dot{\boldsymbol{r}}.\boldsymbol{n} = \ddot{\boldsymbol{r}}.\boldsymbol{n} = 0$. Thus, in (*1*), taking the scalar product of both sides with $\boldsymbol{n}$ yields

$$R = -m\boldsymbol{g}\,.\,\boldsymbol{n}$$

so that (*1*) becomes

$$m\ddot{\boldsymbol{r}} = \boldsymbol{F} + m\boldsymbol{g}', \quad \boldsymbol{g}' = \boldsymbol{g} - (\boldsymbol{g}\,.\,\boldsymbol{n})\boldsymbol{n}.$$

Equation (*3*), differentiated with respect to time, gives an equation for $\dot{\boldsymbol{\omega}}$, which can then be eliminated by combining with equation (*2*). Thus from (*3*),

$$\begin{aligned} \ddot{\boldsymbol{r}} &= -a\boldsymbol{n}\times\dot{\boldsymbol{\omega}}, \\ &= \boldsymbol{n}\times(\boldsymbol{n}\times\boldsymbol{F})a^2/mk^2, \quad \text{using } (2) \\ &= -a^2\boldsymbol{F}/mk^2. \end{aligned}$$

We can now eliminate $\boldsymbol{F}$, using (*1*), giving

$$m\ddot{\boldsymbol{r}} = -mk^2\ddot{\boldsymbol{r}}/a^2 + m\boldsymbol{g}'$$

i.e.

$$\ddot{\boldsymbol{r}} = \boldsymbol{g}'/(1+k^2/a^2).$$

Since the acceleration of the centre of mass is constant, we may immediately conclude that the general motion of C is a parabolic path.

3.3. *A sphere of radius a and radius of gyration k rolls without slipping on the outer surface of a fixed sphere of radius b. Show that this motion is described by the equation*

$$(1+b/a)(1+a^2/k^2)\boldsymbol{n}\times\ddot{\boldsymbol{n}} + \lambda\dot{\boldsymbol{n}} + (ag/k^2)\boldsymbol{n}\times\boldsymbol{z} = 0,$$

where $\boldsymbol{n}$ *is the outward unit vector along the line of centres and* $\boldsymbol{z}$ *is the unit vector along the upward vertical;* λ *is a constant to be identified.*

Show that this motion is equivalent to the motion under gravity of a suitably chosen symmetric top pivoted at a point on its axis distance a from its mass centre.

The same principle applies as in the previous problem, though we shall meet more complications here.

Let the sphere have mass m, centre G. O is a fixed origin. Let $\boldsymbol{\omega}$ be the angular velocity of the sphere. Then relative to O, if G has position vector $\boldsymbol{r}$,

$$m\ddot{\boldsymbol{r}} = \boldsymbol{R} - mg\boldsymbol{z} \qquad \text{linear momentum;}$$

$$mk^2\dot{\boldsymbol{\omega}} = \overrightarrow{GP}\times\boldsymbol{R} \qquad \text{angular momentum;}$$

$$\dot{\boldsymbol{r}} + \boldsymbol{\omega}\times\overrightarrow{GP} = 0 \qquad \text{rolling condition.}$$

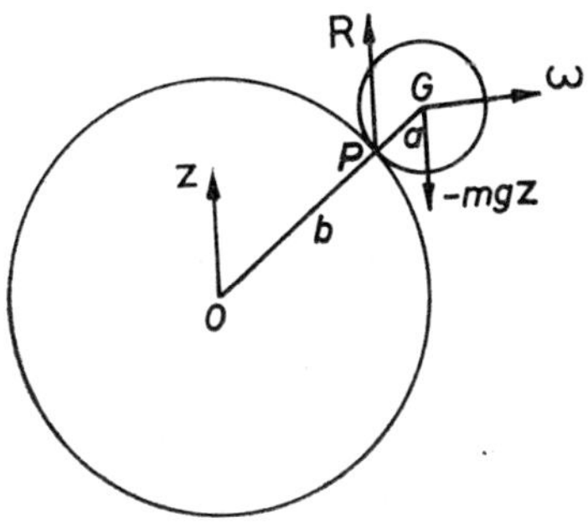

FIG. 16

Now, $\boldsymbol{r} = (a+b)\boldsymbol{n}$. Hence,

$$m(a+b)\ddot{\boldsymbol{n}} = \boldsymbol{R} - mg\boldsymbol{z}, \tag{1}$$

$$mk^2\dot{\boldsymbol{\omega}} = -a\boldsymbol{n}\times\boldsymbol{R}, \tag{2}$$

$$(a+b)\dot{\boldsymbol{n}} + a\boldsymbol{n}\times\boldsymbol{\omega} = 0. \tag{3}$$

We have to eliminate the unwanted quantities $\boldsymbol{\omega}$ and $\boldsymbol{R}$ from the three equations above. From (2),

$$\boldsymbol{n}\,.\,\dot{\boldsymbol{\omega}} = 0$$

From (3),

$$\dot{\boldsymbol{n}}\,.\,\boldsymbol{\omega} = 0$$

Hence,

$$\boldsymbol{n}\,.\,\boldsymbol{\omega} = \text{constant} = c, \text{ say.} \tag{4}$$

Now to eliminate $\boldsymbol{R}$: In (2), there appears $\boldsymbol{n}\times\boldsymbol{R}$. Hence $\boldsymbol{R}$ can be eliminated by first taking the vector product of (1) with $\boldsymbol{n}$, i.e.

$$am(a+b)\boldsymbol{n}\times\ddot{\boldsymbol{n}} = a\boldsymbol{n}\times\boldsymbol{R} - mga\boldsymbol{n}\times\boldsymbol{z}.$$

Hence,

$$a(a+b)\boldsymbol{n}\times\ddot{\boldsymbol{n}} = -k^2\dot{\boldsymbol{\omega}}-ga\boldsymbol{n}\times\boldsymbol{z}. \tag{5}$$

We still have to eliminate $\dot{\boldsymbol{\omega}}$. This can be effected most simply by taking the vector product of (3) with $\boldsymbol{n}$, giving

$$(a+b)\boldsymbol{n}\times\dot{\boldsymbol{n}}+a\boldsymbol{n}\times(\boldsymbol{n}\times\boldsymbol{\omega}) = 0,$$

i.e.

$$(a+b)\boldsymbol{n}\times\dot{\boldsymbol{n}}+a[\boldsymbol{n}(\boldsymbol{n}\,.\,\boldsymbol{\omega})-\boldsymbol{\omega}],$$

i.e.

$$(a+b)\boldsymbol{n}\times\dot{\boldsymbol{n}}+ac\boldsymbol{n}-a\boldsymbol{\omega} = 0, \text{ using } (4).$$

We thus obtain a second equation for $\dot{\boldsymbol{\omega}}$,

$$a\dot{\boldsymbol{\omega}} = (a+b)\boldsymbol{n}\times\ddot{\boldsymbol{n}}+ac\dot{\boldsymbol{n}} \tag{6}$$

Elimination of $\dot{\boldsymbol{\omega}}$ between (5) and (6) yields the required equation

$$(1+b/a)(1+a^2/k^2)\boldsymbol{n}\times\ddot{\boldsymbol{n}}+\lambda\dot{\boldsymbol{n}}+(ga/k^2)\boldsymbol{n}\times\boldsymbol{z} = 0,$$

where it is now clear that $\lambda = c =$ component of $\boldsymbol{\omega}$ along $\boldsymbol{n}$. Physically, this result is expected since there can be no torque on the rolling sphere about the line of centres of the two spheres.

Next, we have to show that the motion of the sphere is dynamically equivalent to that of a suitably chosen symmetric top, acted on by gravity. We consider the general equation of motion of a top whose principal moments of inertia are I, I, I_3, the last being the moment of inertia about the symmetry axis, $\boldsymbol{n}$. The angular velocity $\boldsymbol{\omega}$ of the top can be expressed in the form

$$\boldsymbol{\omega} = (\boldsymbol{\omega}\,.\,\boldsymbol{n})\boldsymbol{n}+[\boldsymbol{\omega}-(\boldsymbol{\omega}\,.\,\boldsymbol{n})\boldsymbol{n}].$$

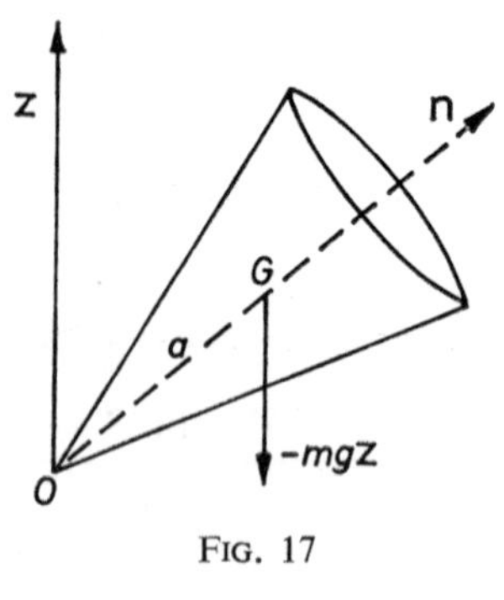

FIG. 17

The terms on the right hand side are, respectively, the components of $\boldsymbol{\omega}$ parallel and perpendicular to $\boldsymbol{n}$. Since

$$\boldsymbol{\omega}-(\boldsymbol{\omega} \, . \, \boldsymbol{n})\boldsymbol{n} = \boldsymbol{n}\times(\boldsymbol{\omega}\times\boldsymbol{n}),$$

and

$$\frac{d\boldsymbol{n}}{dt} = \boldsymbol{\omega}\times\boldsymbol{n},$$

$$\boldsymbol{\omega} = \lambda\boldsymbol{n}+\boldsymbol{n}\times\frac{d\boldsymbol{n}}{dt},$$

where again

$$\lambda = \boldsymbol{\omega} \, . \, \boldsymbol{n}.$$

The angular momentum $\boldsymbol{L}$ about O of the top is clearly

$$\boldsymbol{L} = I_3\lambda\boldsymbol{n}+I\boldsymbol{n}\times\frac{d\boldsymbol{n}}{dt}.$$

We can now obtain an equation of motion for the top

$$\frac{d\boldsymbol{L}}{dt} = \text{external torque about } O,$$

i.e.

$$I_3\frac{d}{dt}(\lambda\boldsymbol{n})+I\boldsymbol{n}\times\ddot{\boldsymbol{n}} = -mga\boldsymbol{n}\times\boldsymbol{z}.$$

To show that λ is a constant of the motion, we note that the scalar product of the equation above with $\boldsymbol{n}$ gives

$$\boldsymbol{n} \, . \, \frac{d}{dt}(\lambda\boldsymbol{n}) = 0.$$

Since $\boldsymbol{n} \, . \, d\boldsymbol{n}/dt = 0$ ($\boldsymbol{n}$ being a unit vector), this condition reduces to

$$\frac{d\lambda}{dt} = 0.$$

Dividing the resulting equation by I_3 gives

$$(I/I_3)\boldsymbol{n}\times\ddot{\boldsymbol{n}}+\lambda\dot{\boldsymbol{n}}+(mga/I_3)\boldsymbol{n}\times\boldsymbol{z} = 0.$$

To obtain equivalence with the equation of the rolling sphere, we require

$$I_3 = mk^2, \quad I = m(1+b/a)(a^2+k^2).$$

3.4. *A sphere of radius a, mass M and radius of gyration k about any axis through its centre, rolls on top of a rough horizontal cylinder, radius b. Let* $\boldsymbol{i}$ *be a unit vector along a line from the point of contact through the centre of the sphere,* $\boldsymbol{j}$ *a unit vector along the axis of the cylinder, so that the position vector* $\boldsymbol{r}$ *of the centre of the sphere with respect to a fixed point on the axis is given by*

$$\boldsymbol{r} = y\boldsymbol{j}+(a+b)\boldsymbol{i}$$

Show that

$$a\dot{n}\boldsymbol{i}+an\frac{d\boldsymbol{i}}{dt}+\dot{y}\frac{d\boldsymbol{i}}{dt}\times\boldsymbol{j}$$

$$+(1+a^2/k^2)\left[\ddot{y}\boldsymbol{i}\times\boldsymbol{j}+(a+b)\boldsymbol{i}\times\frac{d^2\boldsymbol{i}}{dt^2}\right]+ga^2/k^2\boldsymbol{i}\times\boldsymbol{z} = 0,$$

where n is the component of angular velocity along $\boldsymbol{i}$ *and* $\boldsymbol{z}$ *is a unit vector pointing vertically upwards. Hence show that*

$$(a+b)(a^2+k^2)\ddot{\theta} = a^2 g \sin\theta$$

where θ *is the angle between* $\boldsymbol{i}$ *and* $\boldsymbol{z}$.

$$\boldsymbol{r}_c = \boldsymbol{r} = y\boldsymbol{j}+(a+b)\boldsymbol{i}.$$

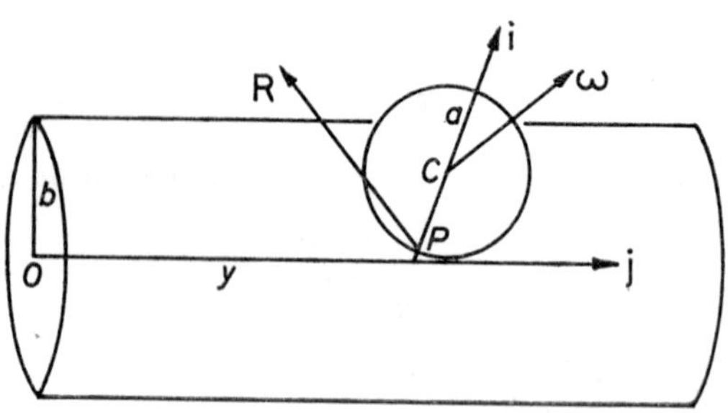

FIG. 18

Linear momentum:

$$M\ddot{\boldsymbol{r}} = -Mg\boldsymbol{z}+\boldsymbol{R}. \tag{1}$$

Angular momentum:

$$Mk^2\dot{\boldsymbol{\omega}} = -a\boldsymbol{i}\times\boldsymbol{R}. \tag{2}$$

Let P be the point of contact. Then the rolling condition is $\dot{\boldsymbol{r}}_P = 0$, i.e.

$$\dot{\boldsymbol{r}}-a\boldsymbol{\omega}\times\boldsymbol{i} = 0. \tag{3}$$

We use $\dot{\boldsymbol{r}} = \dot{y}\boldsymbol{j}+(a+b)(d\boldsymbol{i}/dt)$ in the rolling condition and a similar expression for $\ddot{\boldsymbol{r}}$ in the equation of linear momentum (*1*). The object now is to eliminate the unwanted quantities, the angular velocity $\boldsymbol{\omega}$ and the reaction $\boldsymbol{R}$. Consider the vector product of equation (*1*) with $-a\boldsymbol{i}$. We find

$$-Ma\ddot{y}\boldsymbol{i}\times\boldsymbol{j}-Ma(a+b)\boldsymbol{i}\times\frac{d^2\boldsymbol{i}}{dt^2} = Mga\boldsymbol{i}\times\boldsymbol{z}-a\boldsymbol{i}\times\boldsymbol{R},$$

$$= Mga\boldsymbol{i}\times\boldsymbol{z}+M\dot{\boldsymbol{\omega}}k^2/a.$$

Equation (*3*) can be written as

$$\dot{y}\boldsymbol{j}+(a+b)\frac{d\boldsymbol{i}}{dt} = a\boldsymbol{\omega}\times\boldsymbol{i}.$$

Hence

$$\dot{y}\boldsymbol{i}\times\boldsymbol{j}+(a+b)\boldsymbol{i}\times\frac{d\boldsymbol{i}}{dt} = a\boldsymbol{i}\times(\boldsymbol{\omega}\times\boldsymbol{i}),$$

$$= a(\boldsymbol{\omega}-n\boldsymbol{i}),$$

so

$$a\dot{\boldsymbol{\omega}} = a\dot{n}\boldsymbol{i}+an\frac{d\boldsymbol{i}}{dt}+\ddot{y}\boldsymbol{i}\times\boldsymbol{j}+\dot{y}\frac{d\boldsymbol{i}}{dt}\times\boldsymbol{j}+(a+b)\boldsymbol{i}\times\frac{d^2\boldsymbol{i}}{dt^2}.$$

This enables us to eliminate $\dot{\boldsymbol{\omega}}$ between the equation above and equation (*2*). The eliminant is the required equation.

For the last part of the problem, we note that $\boldsymbol{i}\times\boldsymbol{z} = \boldsymbol{j}\sin\theta$. Hence taking the scalar product of the above eliminant with $\boldsymbol{j}$ and noting that

$$\boldsymbol{i}\,.\,\boldsymbol{j} = 0,\quad \frac{d\boldsymbol{i}}{dt} = -\dot{\theta}\boldsymbol{j}\times\boldsymbol{i},\quad \frac{d^2\boldsymbol{i}}{dt^2} = -\ddot{\theta}\boldsymbol{j}\times\boldsymbol{i}+\dot{\theta}^2\boldsymbol{i},$$

the only terms which remain are

$$\boldsymbol{j}\,.\,[(1+a^2/k^2)(a+b)\boldsymbol{i}\times\ddot{\theta}(\boldsymbol{i}\times\boldsymbol{j})+g(a^2/k^2)\boldsymbol{j}\sin\theta] = 0,$$

i.e.

$$(a+b)(a^2+k^2)\ddot{\theta} = ga^2\sin\theta,$$

3.5. *A uniform disc of radius a rolls and slips on a horizontal plane. Show that*

$$\frac{4}{a}\left(g\boldsymbol{z}+\frac{d^2\boldsymbol{r}}{dt^2}\right)\times\boldsymbol{j} = 2\frac{d}{dt}(n\boldsymbol{i})+\boldsymbol{i}\times\frac{d^2\boldsymbol{i}}{dt^2}$$

where **i**, **j** *and* **z** *are unit vectors,* **i** *normal to the disc,* **j** *pointing upwards along the diameter through the point of contact with the plane, and* **z** *vertically upwards.* **r** *is the position vector of the centre of the disc, and n is the component of angular velocity of the disc along* **i**.

If the disc rolls without slipping and **i** *precesses uniformly with period $2\pi/\omega$ about* **z** *at an angle α to z, show that*

$$4(g/a)\cos\alpha+6n\omega\sin\alpha = \omega^2\cos\alpha\sin\alpha.$$

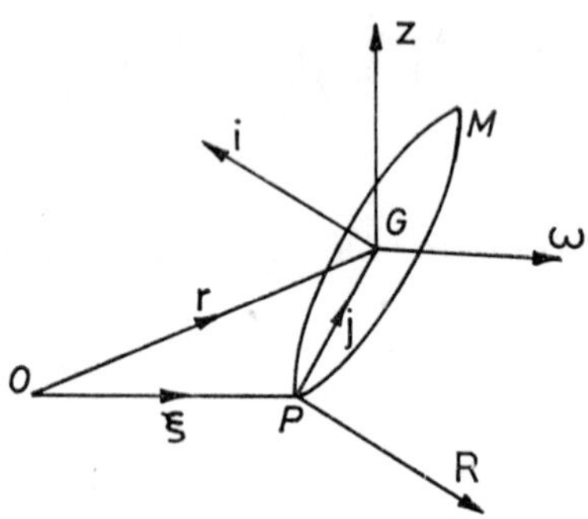

FIG. 19

In Fig. 19, $\boldsymbol{r} = \boldsymbol{\xi}+a\boldsymbol{j}$, $\boldsymbol{R}$ is the reaction at the point of contact and M is the mass of the disc.

Linear momentum:

$$M\frac{d^2\boldsymbol{r}}{dt^2} = \boldsymbol{R}-Mg\boldsymbol{z}. \qquad (1)$$

Angular momentum:

$$\frac{d\boldsymbol{L}_G}{dt} = -a\boldsymbol{j}\times\boldsymbol{R}.$$

$$\begin{aligned}\boldsymbol{L}_G &= \tfrac{1}{2}Ma^2(\boldsymbol{\omega}\,.\,\boldsymbol{i})\boldsymbol{i}+\tfrac{1}{4}Ma^2(\boldsymbol{\omega}-(\boldsymbol{\omega}\,.\,\boldsymbol{i})\boldsymbol{i}),\\ &= \tfrac{1}{2}Ma^2n\boldsymbol{i}+\tfrac{1}{4}Ma^2\boldsymbol{i}\times(\boldsymbol{\omega}\times\boldsymbol{i}),\\ &= \tfrac{1}{2}Ma^2n\boldsymbol{i}+\tfrac{1}{4}Ma^2\boldsymbol{i}\times\frac{d\boldsymbol{i}}{dt}\end{aligned}$$

(using the fact that $\boldsymbol{i}$ rotates with angular velocity $\boldsymbol{\omega}$, since it is fixed normal to the disc). Hence

$$\frac{d\boldsymbol{L}_G}{dt} = \tfrac{1}{2}Ma^2\frac{d}{dt}(n\boldsymbol{i})+\tfrac{1}{4}Ma^2\boldsymbol{i}\times\frac{d^2\boldsymbol{i}}{dt^2},$$

and the equation of angular momentum is

$$\tfrac{1}{2}Ma^2\frac{d}{dt}(n\boldsymbol{i})+\tfrac{1}{4}Ma^2\boldsymbol{i}\times\frac{d^2\boldsymbol{i}}{dt^2} = -a\boldsymbol{j}\times\boldsymbol{R}. \qquad (2)$$

The standard procedure of eliminating $\boldsymbol{R}$ is followed. Thus we form the vector product of (*1*) with $-a\boldsymbol{j}$, producing the term $-a\boldsymbol{R}\times\boldsymbol{j}$ which appears also in (*2*). On combining both equations, we obtain

$$-\frac{4}{a}\frac{d^2\boldsymbol{r}}{dt^2}\times\boldsymbol{j}+2\frac{d}{dt}(n\boldsymbol{i})+\boldsymbol{i}\times\frac{d^2\boldsymbol{i}}{dt^2}-\frac{4g}{a}\boldsymbol{z}\times\boldsymbol{j} = 0. \qquad (3)$$

Apart from a minor rearrangement, this is the result we are seeking.

Notice that no rolling condition has been invoked so far, and that in general slip will occur at the point of contact P. We now have to find the condition for pure rolling and impose it on equation (*3*), together with additional steady state conditions on the motion of the disc.

Rolling condition: This is again the condition that P is instantaneously at rest.

$$\begin{aligned}\boldsymbol{v}_P &= \boldsymbol{v}_G+\boldsymbol{\omega}\times\overrightarrow{GP}\\ &= \dot{\boldsymbol{r}}-a\boldsymbol{\omega}\times\boldsymbol{j} = 0.\end{aligned}$$

Hence

$$\dot{\boldsymbol{r}} = a\boldsymbol{\omega}\times\boldsymbol{j}.$$

Since $\boldsymbol{i}$ precesses uniformly with angular velocity $\omega\boldsymbol{z}$, $d\boldsymbol{i}/dt = \omega\boldsymbol{z}\times\boldsymbol{i}$. Also, $\boldsymbol{z} = \boldsymbol{i}\cos\alpha+\boldsymbol{j}\sin\alpha$, so that $d\boldsymbol{i}/dt = -\omega\sin\alpha\boldsymbol{k}$, where $\boldsymbol{k} = \boldsymbol{i}\times\boldsymbol{j}$. Hence the rolling condition takes the form

$$\begin{aligned}\frac{d\boldsymbol{r}}{dt} &= a\boldsymbol{\omega}\times\boldsymbol{j},\\ &= a\left(n\boldsymbol{i}+\boldsymbol{i}\times\frac{d\boldsymbol{i}}{dt}\right)\times\boldsymbol{j},\\ &= an\boldsymbol{k}, \quad \text{since } (\boldsymbol{i}\times\boldsymbol{k})\times\boldsymbol{j} = 0.\end{aligned}$$

Then $d^2\boldsymbol{r}/dt^2 = an\omega\boldsymbol{z}\times\boldsymbol{k}$, since $\boldsymbol{k}$ also precesses about $\boldsymbol{z}$ with angular velocity ω, and n is now constant. Therefore the first term in equation (*3*) is

$$-4n\omega(\boldsymbol{z}\times\boldsymbol{k})\times\boldsymbol{j} = -4n\omega\boldsymbol{k}\sin\alpha.$$

The second term is

$$2n\frac{d\boldsymbol{i}}{dt} = -2n\omega\boldsymbol{k}\sin\alpha.$$

The third term is

$$\begin{aligned}\boldsymbol{i}\times\frac{d^2\boldsymbol{i}}{dt^2} &= \boldsymbol{i}\times\left(\omega\boldsymbol{z}\times\frac{d\boldsymbol{i}}{dt}\right),\\ &= \boldsymbol{i}\times(\boldsymbol{z}\times\boldsymbol{k})(-\omega^2\sin\alpha),\\ &= \omega^2\boldsymbol{k}\sin\alpha\cos\alpha.\end{aligned}$$

The last term is simply $-4(g/a)\boldsymbol{k}\cos\alpha$.

Putting all these terms together gives for (*3*)

$$4(g/a)\cos\alpha+6n\omega\sin\alpha = \omega^2\sin\alpha\cos\alpha,$$

which is the required result.

EXERCISES

1. A square lamina *ABCD* is rotating freely with angular velocity ω about the diagonal *BD* when suddenly the mid-point of the side *AB* is seized and held fixed. Find the initial magnitude and direction of the angular velocity of the rotation about this point.

2. A uniform sphere rolls without slipping, on a horizontal plane which rotates with a constant angular velocity about a vertical axis. Determine the equation of motion of the centre of the sphere and show that it follows a circular trajectory.

3. Calculate the change in inertia tensor of a rigid body if the point of reference in the body, originally the mass centre G, is displaced through the vector distance R. Hence obtain the inertia tensor of a uniform right circular cylinder of mass M radius a and height $2h$, with respect to a point O on an edge of the cylinder.

The cylinder is constrained to rotate about a fixed axis through O with constant angular velocity $\omega \boldsymbol{n}$, $\boldsymbol{n}$ being a unit vector along the axis. Calculate the angular momentum of the cylinder and the reactive couple on the axis.

4. Establish the relation between the angular momentum of a rigid body about any point O and that about its mass centre.

OO' is the axis of a uniform solid right circular cylinder of height h and radius a. C is a point on the circumference of the plane end with centre O'. The cylinder is turning freely about OC, with angular velocity ω.

By evaluating its components with respect to suitable axes fixed in the cylinder, find the angular momentum of the cylinder about O.

The point O' is suddenly fixed (so that the cylinder now rotates freely about O'). Find the new direction of the instantaneous axis of rotation.

5. A symmetrical gyrostat rotates with spin n about a point O on its axis and precesses uniformly with precessional angular velocity Ω. Show that the moment about O of the forces acting on the gyrostat must be given by

$$N = (Cn\Omega - A\Omega^2 \cos \alpha)\boldsymbol{z} \times \boldsymbol{k}$$

where $\boldsymbol{z}$, $\boldsymbol{k}$ are the unit vectors along the respective axes of precession and of the gyrostat and α is the angle between them. What are C and A?

6. An axially symmetrical rigid body, with principal moments of inertia A, A, C, is rotating about a fixed point O with angular velocity $\boldsymbol{\omega}$. Obtain the equations of motion of the body referred to an orthogonal set of unit vectors $\boldsymbol{i}, \boldsymbol{j}, \boldsymbol{k}$ rotating about O with angular velocity $\boldsymbol{\Omega}$ such that $\boldsymbol{k}$ coincides with the symmetry axis of the body.

A uniform circular disc, centre G, is mounted on an axis through G normal to the disc. The system is mounted on the edge of a horizontal turntable rotating with constant angular velocity Ω. The axis of the disc is constrained so that it can rotate freely about G in a vertical plane only, this vertical plane being fixed tangential to the edge of the turntable. Show that

$$\ddot{\theta}+2\Omega n \sin\theta-\Omega^2 \sin\theta\cos\theta = 0$$

where θ is the angle between the vertical and the axis of the disc and n is a constant of the motion to be identified.

Describe the motion if $\Omega \ll n$.

4 EULER'S EQUATIONS

Euler's equations take advantage of various simplifying aspects of the introduction of rotating axes. In general, the angular momentum $\boldsymbol{L}$ and the angular velocity $\boldsymbol{\omega}$ of a rigid body are not parallel, but are related by an equation

$$\boldsymbol{L} = \boldsymbol{I} \,.\, \boldsymbol{\omega},$$

where $\boldsymbol{I}$ is the inertia tensor. This equation may be regarded as a short form of writing three equations which show that the individual components L_x, L_y and L_z are linear combinations of ω_x, ω_y and ω_z. The coefficients are the moments and products of inertia.

In any general motion of a rigid body which is not spherically symmetric, $\boldsymbol{I}$ will be continuously changing as the orientation of the body changes. The angular momentum equation becomes very complicated. The first step is to recognise that if $\boldsymbol{I}$ is referred to a set of axes fixed in the *body*, $\boldsymbol{I}$ will be constant, and

$$\frac{d\boldsymbol{L}}{dt} = \boldsymbol{I} \,.\, \frac{d\boldsymbol{\omega}}{dt} = \boldsymbol{N}.$$

So far the choice of body axes is quite arbitrary. The second simplification is effected by choosing a *particular* set of body axes, the principal axes. These always exist and are a set of orthogonal axes fixed in the body with respect to which all products of inertia vanish (the inertia tensor is diagonal). A consequence of this is that $\boldsymbol{L}$ and $\boldsymbol{\omega}$ are parallel if and only if the axis of rotation is a principal axis.

Writing out the angular momentum equation in components then yields Euler's equations, as any text book on dynamics will show.

4.1. *An axially symmetric rigid body rotates with angular velocity* $\boldsymbol{\omega}$ *about a fixed point O, on the symmetry axis, with respect to which the principal moments of inertia are* I, I, I_3. *The motion is resisted by a* *frictional torque* $-k\boldsymbol{\omega}$, *k being a constant. Given that the angular velocity is not initially aligned along a principal axis, show that eventually the instantaneous axis of rotation of the body will align itself along or perpendicular to the symmetry axis according as* $I \gtrless I_3$. *What will result if* $I = I_3$*?*

If $\boldsymbol{N} = (N_1, N_2, N_3)$ is the torque on the body about the pivot O, then Euler's equations become

$$I_1\dot{\omega}_1 - (I_2 - I_3)\omega_2\omega_3 = N_1,$$

$$I_2\dot{\omega}_2 - (I_3 - I_1)\omega_3\omega_1 = N_2,$$

$$I_3\dot{\omega}_3 - (I_1 - I_2)\omega_1\omega_2 = N_3,$$

where, in general, the principal moments of inertia are taken as I_1, I_2, I_3. We have now to set $I_1 = I_2 = I$, while $N_1 = -k\omega_1$, and so on. We thus obtain the following set of equations

$$I\dot{\omega}_1 - (I - I_3)\omega_2\omega_3 = -k\omega_1 \qquad (1)$$

$$I\dot{\omega}_2 + (I - I_3)\omega_3\omega_1 = -k\omega_2 \qquad (2)$$

$$I_3\dot{\omega}_3 = -k\omega_3 \qquad (3)$$

Initially, $\boldsymbol{\omega}$ is quite arbitrary. We have to show that as $t \to \infty$, $\omega_3/\omega_\perp \to 0$ or ∞, according as $I \gtrless I_3$, where $\omega_\perp = (\omega_1^2 + \omega_2^2)^{\frac{1}{2}}$. Equation (*3*) yields

$$\omega_3 = A \exp(-kt/I_3).$$

Multiply (*1*) by ω_1, (*2*) by ω_2 and add the resulting equations, obtaining

$$I(\omega_1\dot{\omega}_1 + \omega_2\dot{\omega}_2) = -k(\omega_1^2 + \omega_2^2),$$

that is,

$$I\frac{d}{dt}(\omega_\perp^2) = -2k\omega_\perp^2.$$

(Another way of obtaining the equation above is to use the fact that the rate of change of kinetic energy of rotation of the rigid body $\boldsymbol{\omega} \cdot \boldsymbol{N} = -k\omega^2$.) Hence

$$\omega_\perp = B \exp(-kt/I).$$

A and B are arbitrary (non-zero) constants, being the initial values of ω_3 and ω respectively. Thus,

$$\frac{\omega_3}{\omega_\perp} = \frac{A}{B}\exp\left[-kt\left(\frac{1}{I_3}-\frac{1}{I}\right)\right]$$

$$\to 0 \text{ or } \infty \text{ according as } I_3 < I \text{ or } I_3 > I.$$

If $I = I_3$, the inertia tensor is doubly degenerate and *any* axis is a principal axis (for example, any diameter of a uniform sphere is a principal axis). The body will in this case continue to rotate about its initial direction, gradually slowing down to rest.

4.2. *Show that in the general motion of a lamina about a fixed point under no external forces, the component of angular velocity in the plane of the lamina is conserved.*

Let the principal moments of inertia of the lamina with respect to the fixed point be I_1, I_2 and I_3. Two of the principal axes will be in the plane of the lamina, the third normal to the lamina. If I_3 is the moment of inertia about the latter, then $I_3 = I_1+I_2$. Euler's equations of motion take on a very simple form because of this relation. They are

$$I_1\frac{d\omega_1}{dt} = (I_2-I_3)\omega_2\omega_3 = -I_1\omega_2\omega_3,$$

$$I_2\frac{d\omega_2}{dt} = (I_3-I_1)\omega_3\omega_1 = I_2\omega_3\omega_1.$$

The third equation is not required.

The equations above reduce to

$$\frac{d\omega_1}{dt}+\omega_2\omega_3 = 0, \tag{1}$$

$$\frac{d\omega_2}{dt}-\omega_3\omega_1 = 0. \tag{2}$$

Note that ω_3 is *not* constant, as is the case for a symmetrical rigid body with $I_1 = I_2$. However, multiplying (*1*) by ω_1 and (*2*) by ω_2 and adding the resulting equations, we obtain

$$\omega_1\frac{d\omega_1}{dt}+\omega_2\frac{d\omega_2}{dt} = 0,$$

i.e.

$$\omega_1^2+\omega_2^2 = \text{constant}.$$

This proves the required result, since ω_1 and ω_2 are the components of ω in the plane of the lamina.

In general, there are two constants of motion in the force-free motion of any rigid body, namely the angular momentum $\boldsymbol{L}$ and the kinetic energy T. Notice that $\boldsymbol{L}$ is constant in space, not with respect to axes fixed in the body. However L^2 is invariant to the choice of axes and L^2 and T are the usual constants of the motion, when referred to body axes. What we have proved for the lamina is not an *extra* constant. It can easily be derived directly from L^2 = constant, T = constant.

4.3. *A rigid body is rotating with constant angular velocity, under the action of no forces. Its principal moments of inertia about the point of rotation are all unequal. Determine, by considering a small perturbation about the steady motion, the stability of the rotation.*

Let the principal moments of inertia be I_1, I_2 and I_3, with $I_1 > I_2 > I_3$. We first examine steady rotation about principal axis 1. Let the angular velocity of the body be $\boldsymbol{\omega}_0 = (\omega_0, 0, 0)$. Euler's equations are obviously satisfied with ω_0 constant. In practice, components ω_2 and ω_3 will never be strictly zero, and we ask what will happen if ω_2 and ω_3 are initially small but non-zero. We shall retain only first order quantities in ω_2 and ω_3. The results will therefore be valid only for small departures from the steady rotation. More specifically, let ξ be a small perturbation on ω_0, so that $\omega = \omega_0+\xi$

$$\xi = (\xi_1, \xi_2, \xi_3) \quad \text{and} \quad |\xi| \ll \omega_0.$$

Euler's equations are

$$I_1 \frac{d\omega_1}{dt}-(I_2-I_3)\omega_2\omega_3 = 0, \tag{1}$$

$$I_2 \frac{d\omega_2}{dt}-(I_3-I_1)\omega_3\omega_1 = 0, \tag{2}$$

$$I_3 \frac{d\omega_3}{dt}-(I_1-I_2)\omega_1\omega_2 = 0. \tag{3}$$

To first order in ξ, we have, from

$$(1),\quad I_1\dot{\xi}_1 = 0, \tag{4}$$

$$(2),\quad I_2\dot{\xi}_2-(I_3-I_1)\xi_3\omega_0 = 0, \tag{5}$$

$$(3),\quad I_3\dot{\xi}_3-(I_1-I_2)\xi_2\omega_0 = 0. \tag{6}$$

From equations (*5*) and (*6*), we can easily eliminate ξ_2 or ξ_3; say we eliminate ξ_3, then we obtain an equation for ξ_2

$$\ddot{\xi}_2 = \frac{(I_3-I_1)(I_1-I_2)}{I_2I_3}\,\omega_0^2\xi_2.$$

Since $I_1 > I_2 > I_3$, the coefficient of ξ_2 is negative. Thus ξ_2 will be oscillatory and bounded. We arrive at the same conclusion about ξ_3. ξ_1 is constant, because of equation (*4*). We therefore deduce that motion about this particular axis is stable.

The same analysis is repeated for the other two principal axes. The relevant coefficients are, respectively,

$$\frac{(I_1-I_2)(I_2-I_3)}{I_3I_1}\,\omega_0^2 \quad\text{and}\quad \frac{(I_2-I_3)(I_3-I_1)}{I_1I_2}\,\omega_0^2.$$

Examination of the sign shows that the coefficient is positive for the first and negative for the second. A positive coefficient implies that (initially) perturbations grow exponentially. The conclusion we draw is that motion about axis 2 is unstable, motion about axis 3 is stable. That is, steady motion about the principal axis with intermediate value of moment of inertia is unstable. The motion which actually takes place after the initially exponential growth in the unstable case is quite complex, requiring the general force-free motion of a rigid body. We are not concerned with that here.

4.4. *It is instructive to solve problem* **1.4** *(p. 10) by directly writing down the equations of motion of the system, for although the use of generalised coordinates greatly simplifies the task of thinking carefully about the physics of the problem, the method of generalised coordinates is not always applicable. In any case it is the physical situation which we wish to examine here more closely.*

We first write down Euler's equations for the lamina, with respect to the set of rotating axes ***i***, ***j***, ***k*** which are fixed in the lamina. Let the angular velocity $\boldsymbol{\omega}$ have components $(\omega_1, \omega_2, \omega_3)$ with respect to these axes. Then

$$I\frac{d\omega_1}{dt}+I\omega_2\omega_3 = N_1, \tag{3}$$

$$kI\frac{d\omega_2}{dt}-kI\omega_3\omega_1 = N_2, \tag{4}$$

$$(1+k)I\frac{d\omega_3}{dt}-(1-k)I\omega_1\omega_2 = N_3. \tag{5}$$

On the right we have introduced the components of the reaction $\boldsymbol{N}$ on the lamina. Since the horizontal axis of rotation is smooth, the torque N_1 about it must vanish. The second condition, that which takes into account the smooth vertical axis, is more difficult to obtain. However, if we resolve $\boldsymbol{N}$ along it, then that component must vanish (just as N_1 vanished). The required condition is thus

$$N_2 \sin\theta+N_3\cos\theta = 0.$$

Since we are not concerned with the actual details of the reaction, we simply form (*4*) $\times \sin\theta$, (*5*) $\times \cos\theta$ and add the resulting equations, with the result

$$k\sin\theta\frac{d\omega_2}{dt}-k\omega_3\omega_1\sin\theta+(1+k)\cos\theta\frac{d\omega_3}{dt}-(1-k)\cos\theta\omega_1\omega_2 = 0. \tag{6}$$

Equations (*3*) and (*6*), together with $N_1 = 0$, are the basic equations of motion for the lamina. We must now relate ω_1, ω_2 and ω_3 to the angles θ and ϕ and the angular velocities $\dot{\theta}$ and $\dot{\phi}$. As we have already seen

$$\omega_1 = \dot{\theta},\quad \omega_2 = \dot{\phi}\sin\theta,\quad \omega_3 = \dot{\phi}\cos\theta.$$

Hence (*3*) becomes

$$\ddot{\theta}+\dot{\phi}^2\sin\theta\cos\theta = 0, \tag{7}$$

while (*6*) becomes

$$k\sin\theta(\ddot{\phi}\sin\theta+\dot{\phi}\cos\theta\dot{\theta})-k\dot{\theta}\dot{\phi}\sin\theta\cos\theta$$
$$+(1+k)\cos\theta(\ddot{\phi}\cos\theta-\dot{\phi}\sin\theta\dot{\theta})-(1-k)\dot{\theta}\dot{\phi}\sin\theta\cos\theta = 0,$$

i.e.

$$(k+\cos^2\theta)\ddot{\phi}-2\sin\theta\cos\theta\dot{\phi}\dot{\theta} = 0,$$

which should be immediately recognisable as

$$\frac{d}{dt}[(k+\cos^2\theta)\dot{\phi}] = 0,$$

i.e.

$$\begin{aligned}(k+\cos^2\theta)\dot{\phi} &= \text{constant}\\ &= (k+\tfrac{1}{2})\Omega, \text{ as before.}\end{aligned}$$

Elimination of $\dot{\phi}$ from (7) gives

$$\ddot{\theta}+\sin\theta\cos\theta\left[\frac{(k+\frac{1}{2})\Omega}{k+\cos^2\theta}\right]^2 = 0.$$

Noting that

$$\ddot{\theta} = \frac{d}{d\theta}(\tfrac{1}{2}\dot{\theta})^2$$

and that

$$\frac{d}{d\theta}\frac{1}{k+\cos^2\theta} = \frac{2\sin\theta\cos\theta}{(k+\cos^2\theta)^2},$$

we see that

$$\begin{aligned}\dot{\theta}^2+\frac{(k+\frac{1}{2})^2\Omega^2}{k+\cos^2\theta} &= \text{constant}\\ &= (k+\tfrac{1}{2})\Omega^2,\end{aligned}$$

which is again the required result.

EXERCISES

1. A uniform circular disc of mass m and radius a rotates with constant angular velocity ω about an axle through its centre inclined at angle 45° to its plane. Find the couple exerted by the axle on its bearings.

2. A uniform circular disc supported at its centre O is set spinning with instantaneous angular velocity n about an axis through O making an angle α with the normal to the disc. Discuss the sub-

sequent motion of the disc, showing in particular that the instantaneous axis of rotation precesses in space with angular velocity $n\sqrt{(1+3\cos^2\alpha)}$ about the angular momentum vector.

3. A rectangular lamina $ABCD$, with sides of length $2a$ and $2b$ respectively, rotates about the diagonal AC with constant angular velocity ω. Determine the magnitude and direction of the angular momentum of the lamina and of the torque exerted by the lamina on the axis of rotation.

The diagonal AC is now released and the centre of mass of the lamina is simultaneously fixed. Discuss the resulting motion, showing in particular that the component of the angular velocity in the plane of the lamina remains constant.

4. A uniform rectangular lamina, whose sides are of length a and $2a$ is set rotating, under the action of no external forces, about its mass centre G. If initially the instantaneous axis of rotation is equally inclined to the three principal axes at G, prove that the angle between the instantaneous axis and the normal to the lamina varies between α and β, where

$$2\tan\alpha = \tan\beta = \sqrt{5}.$$

5. A right circular cone of height h and semivertical angle α is rotating freely about its axis with angular velocity ω when a generator is suddenly seized and held fixed. Find the angular velocity of the motion about this generator.

6. A rigid body, freely pivoted at its centre of mass G and under no forces except the reaction at the pivot, has moments of inertia A, $\frac{3}{2}A$, $2A$, about principal axes $G(x, y, z)$ at G. If it is slightly disturbed from a state of rotation about the z-axis with uniform angular velocity n, show that the instantaneous axis describes in the body an elliptic cone with a period $2\pi\sqrt{3}/n$.

If the body, initially at rest, is acted upon by an impulsive couple $(\epsilon H, \epsilon H, H)$, where ϵ is small, show that the greatest inclination of the z-axis to the instantaneous axis in the ensuing motion is approximately $8\epsilon/3$.

PART 2 ELECTRICITY AND MAGNETISM

5 ELECTROSTATICS, MAGNETOSTATICS AND CURRENTS IN EXTENDED MEDIA

Problems in the above subjects are grouped together because the basic differential equations which determine the appropriate potential function are almost identical. Differences arise in boundary conditions but there are many problems for which a solution, say in electrostatics, can be applied immediately to an analogous problem in magnetism or current distribution.

Rationalised M.K.S. units (or S.I. units) will be used throughout. In order to define symbols and to specify the units in which the quantities are expressed, a brief summary of the basic equations in each subject follows.

Electrostatics

Electric field $= \boldsymbol{E}$ volt/metre (Vm^{-1}).

In free space displacement $\boldsymbol{D} = \epsilon_0 \boldsymbol{E}$ coulomb/sq. metre (Cm^{-2}).

ϵ_0 = permittivity of free space $= 8{\cdot}854 \times 10^{-12}$ farad/metre (Fm^{-1}).

In a dielectric

$$\boldsymbol{D} = \epsilon \boldsymbol{E} = \epsilon_0 \boldsymbol{E} + \boldsymbol{P}, \quad \text{where } \epsilon = K \epsilon_0,$$

K = relative permittivity (or dielectric constant) and $\boldsymbol{P}$ = polarisation (Cm^{-2}).

Gauss' theorem:

$$\int \boldsymbol{D} \,.\, d\boldsymbol{S} = Q \quad \text{over any closed surface,}$$

$$\operatorname{div} \boldsymbol{D} = \nabla \,.\, \boldsymbol{D} = \rho, \quad \rho = \text{charge density } (Cm^{-3}).$$

Potential $= V$ volt (V), $\boldsymbol{E} = -\text{grad } V = -\nabla V$.

Poisson's equation $\nabla^2 V = -\rho/\epsilon$ in a uniform medium.

Laplace's equation $\nabla^2 V = 0$ $(\rho = 0)$.

Boundary conditions at the surface of separation of two media:

(*1*) V continuous (or tangential component of $\boldsymbol{E}$ continuous).

(*2*) $\epsilon \dfrac{\partial V}{\partial n}$ continuous (or normal component of $\boldsymbol{D}$ continuous).

Energy density $= \frac{1}{2}\boldsymbol{D} \cdot \boldsymbol{E}$ joule/cubic metre (Jm^{-3}),

$= \frac{1}{2}\epsilon E^2$ in uniform isotropic medium.

Magnetostatics

Magnetic induction $= \boldsymbol{B}$ weber m^{-2} or tesla (Wb m^{-2} or T).

Magnetic field $= \boldsymbol{H}$ ampère m^{-1} (Am^{-1}).

In free space $\boldsymbol{B} = \mu_0 \boldsymbol{H}$, $\mu_0 =$ permeability of free space $= 4\pi \times 10^{-7}$ henry m^{-1} (Hm^{-1}).

In magnetic material $\boldsymbol{B} = \mu \boldsymbol{H} = \mu_0(\boldsymbol{H}+\boldsymbol{I})$, where $\mu = \bar{\mu}\mu_0$, $\bar{\mu} =$ relative permeability, and $\boldsymbol{I} =$ intensity of magnetisation (Am^{-1}).

Over a closed surface $\int \boldsymbol{B} \cdot d\boldsymbol{S} = 0$ and $\nabla \cdot \boldsymbol{B} = 0$. These equations show that the lines of magnetic induction form closed curves.

Magnetostatic potential $= \Omega$ (A), $\boldsymbol{H} = -\nabla\Omega$, $\nabla^2\Omega = 0$ (μ constant).

Boundary conditions:

(*1*) Ω continuous (or H_t continuous),

(*2*) $\mu \dfrac{\partial \Omega}{\partial n}$ continuous (or B_n continuous).

Energy density $= \frac{1}{2}\boldsymbol{B} \cdot \boldsymbol{H}$ (Jm^{-3})

$= \frac{1}{2}\mu H^2$ in uniform isotropic medium.

Steady currents in extended media

Current density $= \boldsymbol{j}$ (Am^{-2}). Conductivity $= \sigma$ ($\Omega^{-1}\,\text{m}^{-1}$).

Ohm's law $\boldsymbol{j} = \sigma \boldsymbol{E}$.

Continuity equation $\nabla . \boldsymbol{j}+\partial\rho/\partial t = 0$, which gives $\nabla . \boldsymbol{j} = 0$, when there is no variation with time.

The total current leaving an electrode of infinite conductivity immersed in a conducting medium

$$I = \int \boldsymbol{j} . d\boldsymbol{S} \quad \text{over the surface of the electrode,}$$

neglecting the cross-section of the supply wire.

We have $\boldsymbol{E} = -\nabla V$ and $\nabla^2 V = 0$ (σ constant).

Boundary conditions:

(*1*) V continuous (or E_t continuous),

(*2*) $\sigma \dfrac{\partial V}{\partial n}$ continuous (or j_n continuous).

The solution of a potential problem consists of finding a potential function which satisfies Laplace's equation and all the boundary conditions. In the electrostatic or current problems the potential of a conductor or electrode is constant over its surface and $\int \boldsymbol{E} . d\boldsymbol{S}$ over the surface is related to the charge or the current. The lines of the electric field (or lines of flow of current) begin on positive charges and end on negative, or go to infinity. In magnetostatics the lines of $\boldsymbol{B}$ form closed curves or go to infinity. The potential must tend to zero at infinity for a finite system of charges or dipoles, but if a uniform field is postulated say in the z-direction we write $V = -Ez+V'$ where V' is finite at infinity. Singularities can occur at a point charge (q) and at a dipole (moment $\boldsymbol{m}$). In these cases we must find a function $V-q/4\pi \epsilon_0 r$ or $V-(\boldsymbol{m} . \boldsymbol{r})/4\pi \epsilon_0 r^3$, which remains finite.

Special methods can be applied to certain types of problem. If the given system of charges or dipoles possesses axial symmetry the potential can be expressed in spherical harmonics

$$\sum_0^\infty (A_n r^n + B_n r^{-(n+1)}) P_n(\cos\theta)$$

where A_n and B_n are constants and $P_n(\cos\theta)$ is the Legendre polynomial of degree n. The constants are then determined from the boundary conditions or from the known variation of the potential with z. If the system has cylindrical symmetry (a two-dimensional distribution) e.g., if it involves cylinders, line charges or linear currents, the potential may be expressed in cylindrical harmonics which include

$$\log r, \quad r^n(A_n \cos n\theta + B_n \sin n\theta), \quad r^{-n}(A_n \cos n\theta + B_n \sin n\theta).$$

In many of the simpler problems it is sufficient to use only a small number of harmonics by virtue of intelligent anticipation. It is useful to remember that a term proportional to $\cos \theta / r^2$ is due to a dipole and a term $(3 \cos^2 \theta - 1)/r^3$ is due to a quadrupole, and so on.

The method of images provides a ready means of obtaining the field distribution when dealing with planes, spheres and cylinders. The student should remember the results for the simpler configurations. He should also keep in mind the *Uniqueness Theorem* which states that when a potential function is found which satisfies all the conditions then that potential function is unique.

5.1. *A point charge q is placed at a distance d from the centre of an earthed conducting sphere of radius a where d > a. Find the charge on the part of the sphere visible from q and show that the ratio of this charge to the charge on the rest of the sphere is*

$$\sqrt{[(d+a)/(d-a)]}.$$

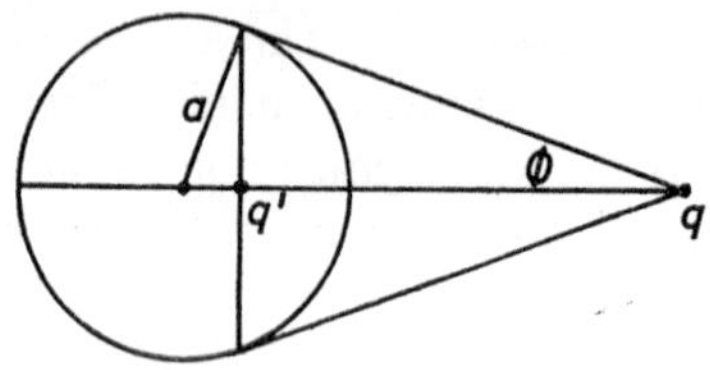

FIG. 20

Consider a point charge q near a surface S. Then, if θ is the angle between $\boldsymbol{E}$ and the normal to the surface at any point, the surface integral of the normal component of $\boldsymbol{E}$ over S is

$$\frac{q}{4\pi\epsilon_0} \int \frac{\cos \theta}{r^2} \, dS = \frac{q\omega}{4\pi\epsilon_0},$$

where ω is the solid angle subtended by S at q.

In the above problem the image of q in the earthed sphere is $q' = -qa/d$ at distance a^2/d from the centre. The charge density on the sphere is $\sigma = \epsilon_0 E_n$ where E_n is the normal component of the resultant electric field due to q and q'. Consider the contributions to E_n due to q and q' separately. The solid angle subtended by the part of the sphere visible to q is

$$2\pi(1-\cos\phi) = 2\pi[1-\sqrt{d^2-a^2}/d].$$

The solid angle subtended by the same surface at q' is 2π, because of the geometry of the figure. Hence the induced charge on this surface is

$$-\epsilon_0\left[\frac{q}{4\pi\epsilon_0}2\pi\left(1-\frac{\sqrt{d^2-a^2}}{d}\right)+\frac{aq}{d}\frac{2\pi}{4\pi\varepsilon_0}\right],$$

$$= -\frac{q}{2}\left[1-\frac{\sqrt{d^2-a^2}}{d}+\frac{a}{d}\right],$$

$$= -\frac{q}{2d}(\sqrt{d+a})(\sqrt{d+a}-\sqrt{d-a}).$$

The charge is negative because E_n is radially inwards.

The charge on the remainder of the sphere may be found by the same method, taking account of the different directions of the contributions to E_n due to q and q', or by using the knowledge that the total charge on the sphere must be $-qa/d$.

Charge on the remainder of the sphere

$$= -q\frac{a}{d}+\frac{q}{2d}(d+a-\sqrt{d^2-a^2}),$$

$$= -\frac{q}{2d}\sqrt{d-a}(\sqrt{d+a}-\sqrt{d-a}).$$

Hence the ratio required $= \sqrt{[(d+a)/(d-a)]}$.

5.2. *Four similar uncharged conducting spheres are placed with their centres at the corners of a square and numbered in rotation. A charge Q is given to sphere* (1) *which is then connected momentarily by a* *thin wire to spheres* (2), (3), (4) *in turn. Show that the final charge on sphere* (1) *is*

$$\frac{Q}{8}\frac{p_{11}-2p_{12}+p_{13}}{p_{11}-p_{12}},$$

where p_{11}, p_{12}, p_{13} *are coefficients of potential.*

From the symmetry of the configuration we can deduce that the potential of any sphere due to unit charge, the other spheres being uncharged, will be the same. Therefore $p_{11} = p_{22} = p_{33} = p_{44}$. For similar reasons $p_{12} = p_{14} = p_{23} = p_{34}$ and $p_{13} = p_{24}$.

We thus require only three coefficients of potential to specify the system and we choose p_{11}, p_{12}, p_{13}.

On connecting (1) to (2) the charge clearly divides equally so that sphere (2) will have charge $Q/2$. On connecting (1) (with charge $Q/2$) to (3) the charge will again, by symmetry, divide equally so that (3) will acquire a charge $Q/4$.

For the last connection, to (4), sphere (1) has a charge $Q/4$ which is shared with (4). Thus

$$V_1 = p_{11}q_1+p_{12}Q/2+p_{13}Q/4+p_{12}q_4,$$
$$V_4 = p_{12}q_1+p_{13}Q/2+p_{12}Q/4+p_{11}q_4,$$

where $V_1 = V_4$ and $q_1+q_4 = Q/4$. We find

$$q_1 = \frac{Q}{8}\frac{p_{11}-2p_{12}+p_{13}}{p_{11}-p_{12}}.$$

5.3. *Two conductors, A and B are such that A is surrounded by B. It is found that when A and B are at potentials V_A and V_B respectively the potential of a point P between them is V_P. The conductors are now earthed and a charge Q is placed at P. Calculate the charges induced on each of the conductors.*

Let us for the identification of the coefficients refer to conductors A and B as (1) and (2) and let us imagine a conductor (3) of infinitesimal dimensions at P. If we give (2) a unit charge with (1) and (3) uncharged then the potentials of all will be the same. Hence

$$p_{22} = p_{12} = p_{32}.$$

We are given that when $V_1 = V_A$, $V_2 = V_B$ then $V_3 = V_P$. Let the charges on A and B under these conditions be q_1 and q_2. Then

$$\left.\begin{aligned} V_A &= p_{11}q_1+p_{12}q_2 \\ V_B &= p_{12}q_1+p_{12}q_2 \\ V_P &= p_{13}q_1+p_{12}q_2 \end{aligned}\right\} \qquad (1)$$

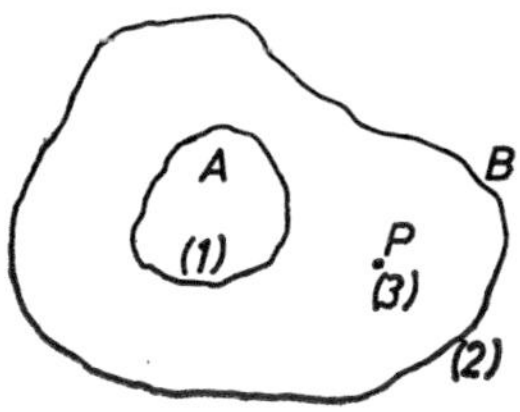

FIG. 21

When A and B are earthed and Q placed at P we have $V_1 = V_2 = 0$ and $q_3 = Q$. Let the charges on A and B now be q_1' and q_2'.

$$0 = p_{11}q_1'+p_{12}q_2'+p_{13}Q,$$
$$0 = p_{12}(q_1'+q_2')+p_{12}Q,$$

i.e.

$$Q = -(q_1'+q_2').$$

This could have been deduced from the fact that the total charge within B must be zero.

$$p_{11}q_1'+p_{12}(-q_1'-Q)+p_{13}Q = 0.$$

Hence

$$q_1' = -Q\frac{p_{13}-p_{12}}{p_{11}-p_{12}}.$$

From equations (*1*)

$$V_P-V_B = (p_{13}-p_{12})q_1, \quad V_A-V_B = (p_{11}-p_{12})q_1,$$

$$q_1' = -Q\frac{V_P-V_B}{V_A-V_B},$$

and

$$q_2' = -(Q+q_1') = -Q\frac{V_A-V_P}{V_A-V_B}.$$

5.4. *A spherical condenser consists of an inner conducting sphere of radius a surrounded by an outer concentric spherical conducting shell of inner radius b and outer radius c. The dielectric constant of the medium between the shells is K_1 and that of the infinite surrounding medium is K. Derive expressions for the potentials of the conductors when they are both insulated and have charges Q_1 and Q_2 respectively.*

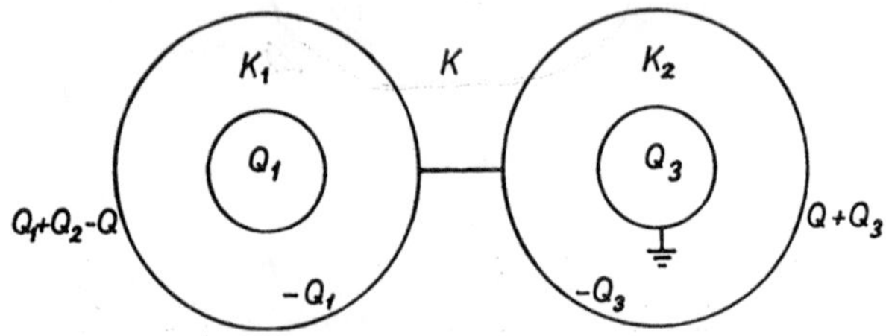

FIG. 22

A second spherical condenser, identical to the first except that the dielectric constant of the medium inside is K_2, is placed at a large distance from the first and is initially uncharged. The inner conductor of this second condenser is now connected to earth and the outer conductor is connected to the outer conductor of the first condenser by a wire of negligible capacity. Show that the loss of energy is

$$\frac{Q^2}{8\pi\epsilon_0 Kc}\left(\frac{2Kc(b-a)+K_2ab}{Kc(b-a)+K_2ab}\right),$$

where Q is the charge that flows along the wire.

A charge $-Q_1$ will be induced on the inner surface of the outer sphere and a charge Q_1+Q_2 will appear on the outer.

Thus

$$V_2 = \frac{1}{4\pi\epsilon_0}\frac{Q_1+Q_2}{Kc}.$$

The potential difference between the spheres

$$V_1 - V_2 = \int_a^b E\,dr = \frac{Q_1}{4\pi\epsilon_0 K_1}\int_a^b \frac{dr}{r^2}$$

$$= \frac{1}{4\pi\epsilon_0}\cdot\frac{Q_1}{K_1}\cdot\frac{b-a}{ab}$$

$$V_1 = \frac{1}{4\pi\epsilon_0}\left(\frac{Q_1+Q_2}{Kc}+\frac{Q_1}{K_1}\frac{b-a}{ab}\right).$$

Let the inner and outer conductors of the second condenser be denoted conductors (3) and (4) respectively. On connection the two outer conductors will have a common potential V and a charge Q_3 will be induced on (3). The final charge distribution is shown in Fig. 22

$$V = \frac{1}{4\pi\Sigma_0}\frac{Q+Q_3}{Kc}$$

for the second condenser.

Since $V_3 = 0$ we have

$$\frac{Q+Q_3}{Kc}+\frac{Q_3}{K_2}\frac{b-a}{ab} = 0,$$

i.e.

$$Q_3 = -Q\frac{K_2 ab}{Kc(b-a)+K_2 ab}.$$

Hence

$$V = \frac{1}{4\pi\epsilon_0}\frac{Q}{Kc}\frac{Kc(b-a)}{Kc(b-a)+Kab} = \frac{Q}{4\pi\epsilon_0}\frac{b-a}{Kc(b-a)+K_2 ab}. \quad (1)$$

Also for the first condenser

$$V = \frac{1}{4\pi\epsilon_0}\cdot\frac{Q_1+Q_2-Q}{Kc}, \quad (2)$$

so

$$V_2 - V = \frac{1}{4\pi\epsilon_0}\frac{Q}{Kc}.$$

The new potential of the inner conductor of the first condenser will be given by

$$V_1' = \frac{1}{4\pi\epsilon_0}\left(-\frac{Q_1+Q_2-Q}{Kc}+\frac{Q_1}{K_1}\cdot\frac{b-a}{ab}\right),$$

$$= V_1 - \frac{1}{4\pi\epsilon_0}\frac{Q}{Kc},$$

giving

$$V_1 - V_1' = \frac{1}{4\pi\epsilon_0}\frac{Q}{Kc}.$$

Using the expression $\frac{1}{2}\Sigma QV$ for the energy of a system of conductors, the loss of energy will be given by

$$[\tfrac{1}{2}Q_1(V_1-V_1')]+[\tfrac{1}{2}Q_2V_2-\tfrac{1}{2}(Q_2-Q)V]-\tfrac{1}{2}QV,$$

$$= \tfrac{1}{2}Q_1(V_1-V_1')+\tfrac{1}{2}Q_2(V_2-V) = \frac{1}{4\pi\epsilon_0}\frac{Q}{2Kc}(Q_1+Q_2).$$

From (*1*) and (*2*)

$$Q_1+Q_2 = Q\left[\frac{2Kc(b-a)+K_2ab}{Kc(b-a)+K_2ab}\right].$$

Hence

$$\text{loss} = \frac{Q^2}{8\pi\epsilon_0 Kc}\left[\frac{2Kc(b-a)+K_2ab}{Kc(b-a)+K_2ab}\right].$$

5.5. *A dielectric sphere of radius R and dielectric constant K has a total charge Q distributed throughout its volume in such a way that the polarisation* **P** *is radial and proportional to* $1/r$, *where r is the distance from the centre* ($r < R$). *Calculate the volume distribution of free charge, the volume and surface distribution of bound charge and the total energy of the system.*

Let

$$\boldsymbol{P} = \frac{\lambda}{r}\hat{\boldsymbol{r}}.$$

The displacement

$$\boldsymbol{D} = \epsilon_0\boldsymbol{E}+\boldsymbol{P} = \epsilon_0 K\boldsymbol{E}.$$

Since $\boldsymbol{D}$, $\boldsymbol{E}$ and $\boldsymbol{P}$ are in the same direction for an isotropic uniform dielectric (assumed here) we have

$$\boldsymbol{D} = \frac{K}{K-1}\boldsymbol{P} = \frac{K}{K-1}\frac{\lambda}{r}\hat{\boldsymbol{r}}.$$

The density of free charge

$$\rho = \nabla \cdot \boldsymbol{D} = \frac{1}{r^2}\frac{\partial}{\partial r}(r^2 D) = \frac{K}{K-1}\frac{\lambda}{r^2}.$$

The total free charge

$$Q = \frac{K\lambda}{K-1}\int_0^R 4\pi r^2\left(\frac{1}{r^2}\right)dr = \frac{K\lambda}{K-1}(4\pi R),$$

giving

$$\lambda = \frac{K-1}{K}\frac{Q}{4\pi R}.$$

Thus

$$D = \frac{Q}{4\pi R}\left(\frac{1}{r}\right) \quad \text{and} \quad \rho = \frac{Q}{4\pi R}\left(\frac{1}{r^2}\right).$$

Volume density of bound charge,

$$\rho' = -\nabla \cdot \boldsymbol{P} = -\frac{K-1}{K}\rho = -\left(\frac{K-1}{K}\right)\left(\frac{Q}{4\pi R}\right)\frac{1}{r^2}.$$

Surface density of bound charge,

$$\sigma' = \boldsymbol{P} \cdot \boldsymbol{n} = \lambda/R = \left(\frac{K-1}{K}\right)\frac{Q}{4\pi R^2}.$$

Total energy of system $W = \frac{1}{2}\int \boldsymbol{D} \cdot \boldsymbol{E}\, d\tau$ over all space. Inside sphere

$$W_1 = \frac{1}{2\epsilon_0 K}\int D^2\, d\tau = \frac{1}{2\epsilon_0 K}\left(\frac{Q}{4\pi R}\right)^2\int_0^R 4\pi\, dr,$$

$$= \frac{Q^2}{4\pi\epsilon_0}\left(\frac{1}{2KR}\right).$$

Outside sphere by Gauss' Theorem $D = Q/4\pi r^2$. Energy outside

$$W_2 = \frac{1}{2\epsilon_0 K}\left(\frac{Q}{4\pi}\right)^2 \int_0^{\infty} \frac{4\pi}{r^2}\,dr = \frac{Q^2}{4\pi\epsilon_0}\left(\frac{1}{2R}\right).$$

Total energy

$$W = W_1 + W_2 = \frac{Q^2}{4\pi\epsilon_0}\left(\frac{K+1}{2KR}\right).$$

5.6. *Find the energy associated with a total charge Q distributed with uniform density throughout the volume of a sphere of radius a.*

The energy associated with a system of charges $q_1 \ldots q_n$ of potential $V_1 \ldots V_n$ is

$$W = \tfrac{1}{2}\sum_{k=1}^{n} q_k V_k.$$

For a distributed charge this becomes $W = \frac{1}{2}\int \rho V\,d\tau$ over the volume occupied by the charge.

The energy can also be found by considering the energy distributed with a density $\frac{1}{2}\mathbf{D}\,.\,\mathbf{E}$.

Thus $W = \frac{1}{2}\int \mathbf{D}\,.\,\mathbf{E}\,d\tau$ over all space.

This becomes $W = \frac{1}{2}\int \epsilon E^2\,d\tau$ for a uniform isotropic medium. We shall solve the problem by both methods.

(*1*) Let ρ = density of charge inside sphere; $Q = \frac{4}{3}\pi a^3 \rho$.

Consider a shell, thickness dr, distant r from centre. The charge on the shell is $4\pi r^2 \rho\,dr$. Potential of shell

$$V = \int_r^{\infty} \mathbf{E}\,.\,\mathbf{dr} = \frac{Q}{4\pi\epsilon_0 a} + \int_r^a E\,dr.$$

Since a spherical shell of charge exerts no field within it the field at distance r from the centre is due only to the charge within the sphere radius r

$$E = \frac{1}{4\pi\epsilon_0}\left(\tfrac{4}{3}\pi r^3 \rho\right)\frac{1}{r^2} = \frac{\rho}{3\epsilon_0}\,r\,.$$

$$\int_r^a E\,dr = \frac{\rho}{3\epsilon_0}\int_r^a r\,dr = \frac{\rho}{6\epsilon_0}(a^2 - r^2).$$

Thus

$$V = \frac{Q}{4\pi\epsilon_0 a} + \frac{\rho}{6\epsilon_0}(a^2 - r^2) = \frac{\rho}{6\epsilon_0}(3a^2 - r^2).$$

$$W = \tfrac{1}{2}\int \rho V \, d\tau = \tfrac{1}{2}\int_0^a \frac{\rho^2}{6\epsilon_0}(3a^2 - r^2)(4\pi r^2 \, dr),$$

$$= \frac{\pi\rho^2}{3\epsilon_0}\int_0^a r^2(3a^2 - r^2)\, dr = \frac{\pi\rho^2}{3\epsilon_0}\left(\frac{4a^5}{5}\right),$$

$$= \frac{Q^2}{4\pi\epsilon_0 a}\left(\frac{3}{5}\right).$$

(2)

$$W = \frac{\epsilon_0}{2}\int E^2 \, d\tau$$

over all space.

Inside the sphere

$$W_1 = \frac{\epsilon_0}{2}\int_0^a \left(\frac{\rho r}{3\epsilon_0}\right)^2 (4\pi r^2 \, dr),$$

$$= \frac{\rho^2}{18\epsilon_0}\left(\frac{4\pi a^5}{5}\right),$$

$$= \frac{Q^2}{4\pi\epsilon_0 a}\left(\frac{1}{10}\right).$$

Outside the sphere

$$W_2 = \frac{\epsilon_0}{2}\int_a^\infty \left(\frac{Q}{4\pi\epsilon_0}\frac{1}{r^2}\right)^2 (4\pi r^2 \, dr),$$

$$= \frac{Q^2}{4\pi\epsilon_0 a}\left(\frac{1}{2}\right).$$

Hence

$$W = W_1 + W_2 = \frac{Q^2}{4\pi\epsilon_0 a}\left(\frac{3}{5}\right).$$

5.7. *A magnetic dipole of moment* **m** *is at the centre of a spherical shell bounded by spheres of radii a and b of material of relative permeability $\bar{\mu}$. Find the potential at any point outside the shell.*

The problem clearly possesses axial symmetry. We have three regions to consider. (1) Within the hollow of the shell we have the given dipole and an additional field which must not include inverse powers of r.

Let

$$\Omega_1 = \frac{m\cos\theta}{4\pi r^2} + \sum_0^\infty A_n r^n P_n(\cos\theta).$$

(2) Within the material of the shell we can have both types of spherical harmonic, i.e.

$$\Omega_2 = \sum_0^\infty (A_n' r^n + B_n' r^{-(n+1)}) P_n(\cos\theta).$$

(3) The distribution of induced magnetism produced by the dipole in the shell must be symmetrical about the plane through the centre normal to the dipole. As a whole, therefore, the distribution of magnetism has a dipole configuration. The potential at a large distance will be that of a dipole. These considerations lead us to try

$$\Omega_3 = \frac{m_3\cos\theta}{4\pi r^2}.$$

It is clear that the continuity of Ω at $r = a$ and $r = b$ simplify the expressions for Ω_1 and Ω_2 above. We need only retain the constants A_1, A_1', B_1', all other constants being zero.

Thus we try

$$\Omega_1 = \frac{m\cos\theta}{4\pi r^2} + A_1 r\cos\theta$$

$$\Omega_2 = \frac{B_1'\cos\theta}{r^2} + A_1' r\cos\theta$$

$$\Omega_3 = \frac{m_3\cos\theta}{4\pi r^2}.$$

Applying the boundary conditions, Ω and $\bar{\mu}(\partial\Omega/\partial n)$ continuous at the two surfaces of the shell, we have

$$r = a. \qquad \frac{m}{4\pi a^2}+A_1 a = \frac{B_1'}{a^2}+A_1' a. \tag{1}$$

$$-\frac{2m}{4\pi a^3}+A_1 = \bar{\mu}\left(-\frac{2B_1'}{a^3}+A_1'\right). \tag{2}$$

$$r = b. \qquad \frac{B_1'}{b^2}+A_1' b = \frac{m_3}{4\pi b^2}. \tag{3}$$

$$\bar{\mu}\left(-\frac{B_1'}{b^3}+A_1'\right) = -\frac{2m_3}{4\pi b^2}. \tag{4}$$

From (3) and (4)

$$3\bar{\mu}B_1' = (\bar{\mu}+2)m_3/4\pi,$$

and

$$A_1' = \frac{m_3}{4\pi b^3}\,\frac{2(\bar{\mu}-1)}{3\bar{\mu}}.$$

Eliminating A_1 from equations (1) and (2) and substituting the values above for A_1' and B_1' we obtain

$$m_3 = \frac{9\bar{\mu}b^3 m}{(\bar{\mu}+2)(2\bar{\mu}+1)b^3-2(\bar{\mu}-1)^2a^3}.$$

The field outside the shell will be the same as that due to a dipole of moment m_3 at the origin with the shell removed.

The field within the magnetic material will be the same as that due to a dipole $4\pi B_1'$ at the centre with the addition of a constant field $-A_1'$ parallel to $\boldsymbol{m}$.

The field within the cavity will be that due to the given dipole $\boldsymbol{m}$ with the addition of a constant field $-A_1$ parallel to $\boldsymbol{m}$. The values of A_1, A_1' and B_1' can be found from the equations.

5.8. *A non-conducting plane lamina, bounded by two concentric circles of radii b and c ($b < c$), is charged with electricity to a uniform surface density σ, and made to rotate in its own plane with constant angular velocity ω about its centre. A soft iron sphere of radius a ($<b$) and relative permeability $\bar{\mu}$ is placed with its centre at the centre of the*

lamina. Find the magnetic field at the centre of the sphere, and show that, at a great distance from the sphere, the field of the sphere is the same as that of a dipole of moment

$$2\pi a^3\sigma\omega(c-b)(\bar{\mu}-1)/(\bar{\mu}+2).$$

Since the problem is symmetrical about the axis of rotation of the lamina we can represent the potential within the iron sphere by

$$\Omega_1 = \sum_0^\infty A_n r^n P_n(\cos\theta).$$

Outside the sphere the potential will be made up of two parts, that due to the rotating lamina and that due to the induced magnetisation of the sphere. The potential due to the lamina at a point on the z-axis can be found as follows.

Consider a ring of width dr. This ring carries a charge $\sigma 2\pi r\,dr$ past any point on it $\omega/2\pi$ times per second. Thus it is equivalent to a circular circuit carrying a current $\omega\sigma r\,dr$. If ϕ is the angle of the cone drawn from a point on the z-axis to the ring the potential due to the current is

$$\frac{\omega\sigma r\,dr}{4\pi}2\pi(1-\cos\phi).$$

The total potential is thus given by

$$\frac{\omega\sigma}{2}\int_b^c r(1-\cos\phi)\,dr = \frac{\omega\sigma}{2}\int_b^c r\left(1-\frac{z}{\sqrt{(r^2+z^2)}}\right)dr$$
$$= \frac{\omega\sigma}{2}\left(\frac{c^2-b^2}{2}-z\sqrt{(c^2+z^2)}+z\sqrt{(b^2+z^2)}\right).$$

For points on the axis where $z < b$ we can expand this expression in powers of z, obtaining

$$\frac{\omega\sigma}{2}\left[\frac{c^2-b^2}{2}-z(c-b)+\frac{z^3}{2}\left(\frac{1}{b}-\frac{1}{c}\right)-\frac{z^5}{8}\left(\frac{1}{b^3}-\frac{1}{c^3}\right)\dots\right].$$

For all points for which $r < b$ the potential due to the rotating lamina can be written

$$\sum_0^\infty A_n' r^n P^n(\cos\theta),$$

which for $\theta = 0$ becomes

$$\sum_0^\infty A_n' z^n.$$

Thus, by comparison with the above expression

$$A_0' = \frac{\omega\sigma}{4}(c^2 - b^2), \quad A_1' = -\frac{\omega\sigma}{2}(c-b), \quad A_2' = 0, \text{ etc.}$$

The potential produced by the sphere at all points outside can be written

$$\sum_0^\infty \frac{B_n}{r^{n+1}} P_n(\cos\theta)$$

since it must tend to zero at infinity.

We thus have, for $r < b$

$$\Omega_1 = \sum_0^\infty A_n r^n P_n(\cos\theta).$$

$$\Omega_2 = \sum_0^\infty A_n' r^n P_n(\cos\theta) + \sum_0^\infty \frac{B_n}{r^{n+1}} P_n(\cos\theta).$$

Applying the boundary conditions at the surface of the sphere $(r = a)$

$$A_n a^n = A_n' a^n + \frac{B_n}{a^{n+1}}, \quad \bar{\mu} A_n n a^{n-1} = A_n' n a^{n-1} - (n+1)\frac{B_n}{a^{n+2}},$$

or

$$A_n = A_n' + \frac{B_n}{a^{2n+1}}, \quad \bar{\mu} A_n = A_n' - \frac{n+1}{n}\frac{B_n}{a^{2n+1}},$$

from which all the constants A_n, B_n can be found in terms of A_n' above.

Here we are primarily interested in A_1 and B_1.

$$A_1 = A_1' + B_1/a^3, \quad \bar{\mu} A_1 = A_1' - 2B/a^3,$$

from which

$$A_1 = \frac{3}{\bar{\mu}+2} A_1' = -\frac{3\omega\sigma(c-b)}{2(\bar{\mu}+2)},$$

$$B_1 = -\frac{a^3(\bar{\mu}-1)}{\bar{\mu}+2} A_1' = \frac{a^3\omega\sigma(c-b)(\bar{\mu}-1)}{2(\bar{\mu}+2)}.$$

The field at the centre of the sphere $= -(\partial\Omega/\partial r)$ for $r = 0$, $\theta = 0$,

$$= -A_1 = \frac{3\omega\sigma(c-b)}{2(\bar{\mu}+2)}.$$

At a great distance the potential due to the sphere will be given approximately by $B_1 \cos\theta/r^2$ since the next term varies as r^{-4}. Thus the equivalent dipole at the centre has a moment

$$4\pi B_1 = 2\pi a^3\omega\sigma(c-b)(\bar{\mu}-1)/(\bar{\mu}+2).$$

5.9. *A parallel-sided air gap of width 2a exists in a large block of iron of relative permeability $\bar{\mu}$. Two long straight parallel wires of negligible cross section carrying current i in opposite directions are placed in contact with the surfaces of the gap opposite to each other. Calculate the magnetic field at a point P at the centre of the gap in the plane of the wires.*

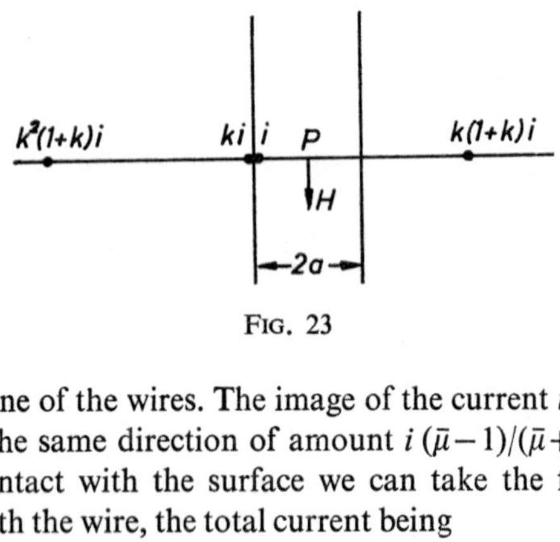

FIG. 23

Consider one of the wires. The image of the current i in the iron is a current in the same direction of amount $i\,(\bar{\mu}-1)/(\bar{\mu}+1)$. Since the wire is in contact with the surface we can take the first image as coincident with the wire, the total current being

$$\left(1+\frac{\bar{\mu}-1}{\bar{\mu}+1}\right)i = (1+k)i,$$

if we write

$$k = \frac{\bar{\mu}-1}{\bar{\mu}+1}.$$

This current will have an image in the opposite face of the gap of amount $k(1+k)i$ situated at a distance of $3a$ from P. This in turn will have an image $k^2(1+k)i$ in the first surface at distance $5a$ from P. There will thus be an infinite series of images due to each wire but since the currents are in opposite directions both series will produce the same field at P.

Hence

$$H = 2\left[(1+k)\frac{i}{2\pi a} - (1+k)k\frac{i}{2\pi(3a)} + (1+k)k^2\frac{i}{2\pi(5a)} - \dots\right]$$

$$= \frac{i}{\pi a}(1+k)\left(1 - \frac{k}{3} + \frac{k^2}{5} - \frac{k^3}{7} + \dots\right).$$

We have

$$\tan^{-1} x = x - \frac{x^3}{3} + \frac{x^5}{5} - \frac{x^7}{7} \dots \ (x \leqq 1),$$

so

$$\tan^{-1}(\sqrt{k}) = \sqrt{k}\left(1 - \frac{k}{3} + \frac{k^2}{5} - \frac{k^3}{7} \dots\right).$$

Hence

$$H = \frac{i}{\pi a}\frac{1+k}{\sqrt{k}}\tan^{-1}\sqrt{k},$$

$$= \frac{i}{\pi a}\frac{2\bar{\mu}}{\sqrt{(\bar{\mu}^2 - 1)}}\tan^{-1}\left(\sqrt{\frac{\bar{\mu}-1}{\bar{\mu}+1}}\right).$$

If $\bar{\mu} \gg 1$

$$H = \frac{i}{\pi a}(2)\frac{\pi}{4} = \frac{i}{2a}.$$

5.10. *Show that, if a magnetic field distribution is due only to magnetised matter,* $\int \boldsymbol{B} \, . \, \boldsymbol{H} \, d\tau = 0$, *where the integral extends over all space.*

Verify this result if the field is that of a uniformly magnetised sphere.

The lines of magnetic induction form closed curves. We can, therefore, divide all space into tubes of induction. Let $\boldsymbol{dl}$ be a short length of one tube and $\boldsymbol{dS}_1$, $\boldsymbol{dS}_2$ the areas of cross-section of its ends. Then since $\int \boldsymbol{B} \, . \, \boldsymbol{dS} = 0$ over any closed surface $\boldsymbol{B}_1 \, . \, \boldsymbol{dS}_1 = \boldsymbol{B}_2 \, . \, \boldsymbol{dS}_2$

or $\boldsymbol{B} \cdot \boldsymbol{dS}$ has a constant value round the tube. The volume of the tube element $= \boldsymbol{dS} \cdot \boldsymbol{dl}$, which we take as $d\tau$. Then, round the tube, $\int \boldsymbol{B} \cdot \boldsymbol{H} \, d\tau = \int (\boldsymbol{B} \cdot \boldsymbol{H})(\boldsymbol{dS} \cdot \boldsymbol{dl})$. Since $\boldsymbol{B}$ and $\boldsymbol{dl}$ are necessarily in the same direction

$$\int \boldsymbol{B} \cdot \boldsymbol{H} \, d\tau = \int (\boldsymbol{B} \cdot \boldsymbol{dS})(\boldsymbol{H} \cdot \boldsymbol{dl}) = (\boldsymbol{B} \cdot \boldsymbol{dS}) \oint \boldsymbol{H} \cdot \boldsymbol{dl}.$$

$\oint \boldsymbol{H} \cdot \boldsymbol{dl}$ round any closed curve is equal to the current flowing through the curve. This is true whether magnetised material is present or not. In this case $\oint \boldsymbol{H} \cdot \boldsymbol{dl} = 0$. The above integral is therefore zero and the same will be true for all the tubes of induction.

Hence for all space

$$\int \boldsymbol{B} \cdot \boldsymbol{H} \, d\tau = 0.$$

For a uniformly magnetised sphere of intensity of magnetisation $\boldsymbol{I}$ the field inside is uniform and

$$\boldsymbol{B} = \tfrac{2}{3}\mu_0 \boldsymbol{I}, \quad \boldsymbol{H} = -\frac{\boldsymbol{I}}{3},$$

i.e. $\boldsymbol{H}$ is opposite in direction to $\boldsymbol{B}$.

Hence within the sphere

$$\begin{aligned} \int \boldsymbol{B} \cdot \boldsymbol{H} \, d\tau &= -\tfrac{2}{9}\mu_0 I^2 (\tfrac{4}{3}\pi a^3). \\ &= -\frac{8\pi\mu_0 I^2 a^3}{27}. \end{aligned}$$

Outside the sphere the field is that of a dipole $\boldsymbol{m} = \tfrac{4}{3}\pi a^3 \boldsymbol{I}$ situated at the centre.

The potential

$$\Omega = \frac{m \cos\theta}{4\pi r^2}$$

where θ is the angle between $\boldsymbol{I}$ and $\boldsymbol{r}$.

$$H_r = -\frac{\partial\Omega}{\partial r} = \frac{2m\cos\theta}{4\pi r^3}.$$

$$H_\theta = -\frac{1}{r}\frac{\partial\Omega}{\partial\theta} = \frac{m\sin\theta}{4\pi r^3}.$$

Thus

$$H^2 = \left(\frac{m}{4\pi r^3}\right)^2 (3\cos^2\theta + 1).$$

Since, outside, $\boldsymbol{B} = \mu_0 \boldsymbol{H}$,

$$\int \boldsymbol{B}\,.\,\boldsymbol{H}\,d\tau = \mu_0 \int H^2\,d\tau \quad \text{where} \quad d\tau = r^2 \sin\theta\,dr\,d\theta\,d\phi.$$

$$= \mu_0 \left(\frac{m}{4\pi}\right)^2 \int_0^{2\pi} d\phi \int_0^{\pi} (3\cos^2\theta + 1)\sin\theta\,d\theta \int_a^{\infty} \frac{dr}{r^4}.$$

$$= \mu_0 \left(\frac{m}{4\pi}\right)^2 (2\pi)(4)\left(\frac{1}{3a^3}\right).$$

$$= \mu_0 \left(\frac{Ia^3}{3}\right)^2 \frac{8\pi}{3a^3}.$$

$$= \frac{8\pi\mu_0 I^2 a^3}{27}.$$

The sum of the two integrals is thus zero.

5.11. *A and B are the opposite ends of a diameter of a thin spherical shell of radius a, thickness t, conductivity σ. Current enters and leaves by two small circular electrodes of radius c centred on A and B. Find the resistance between the electrodes.*

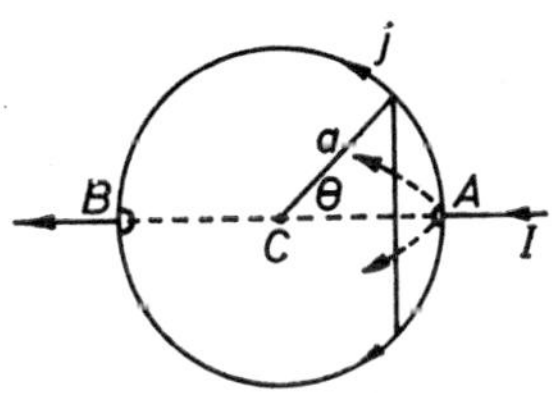

FIG. 24

By symmetry the current will flow along great circles of the sphere, and the current density j will be a function of θ. The total current

$$I = 2\pi a \sin\theta t j,$$

i.e.

$$j = I/2\pi a t \sin\theta.$$

Since $j = \sigma E$ we have

$$E = \frac{I}{2\pi a\sigma t}\frac{1}{\sin\theta}.$$

The potential difference V between A and B will be given by the work done by the field in carrying unit charge from A to B along a line of flow.

$$V = \int_B^A Ea\, d\theta = \frac{I}{2\pi\sigma t}\int_B^A \frac{d\theta}{\sin\theta}.$$

The limits of the integral will be c/a and $\pi - c/a$ approximately.

Hence

$$V = \frac{I}{2\pi\sigma t}\left[\log\tan\frac{\theta}{2}\right]_{c/a}^{\pi - c/a} = \frac{I}{\pi\sigma t}\log\left(\cot\frac{c}{2a}\right).$$

and the resistance

$$R = \frac{V}{I} = \frac{1}{\pi\sigma t}\log\left(\cot\frac{c}{2a}\right).$$

5.12. *Find the resistance between two small circular electrodes of radius c situated a distance 2d apart in an infinite plate of thickness t, conductivity σ.*

The field distribution in this problem is the same as in the two-dimensional field produced by two infinite parallel wires charged $+\rho$ and $-\rho$, per unit length. The plate may be regarded as a slice of an infinite medium. The lines of the electric field (or the lines of flow) are circles centred on the perpendicular bisector of the line joining the electrodes. The electric field due to an infinite wire $= (\rho/2\pi\epsilon_0 r)\hat{\mathbf{r}}$, which may be written $(\lambda/r)\hat{\mathbf{r}}$ for convenience.

At the point P on the perpendicular bisector the resultant field $E = 2\lambda\, d/r^2$ parallel to AB.

Therefore the current density $j = 2\lambda\, d\sigma/r^2$.

The total current

$$I = 2\lambda\, d\sigma t \textstyle\int_{-\infty}^{\infty} dy/r^2,$$

$$= 2\lambda\, d\sigma t\int_{-\infty}^{\infty}\frac{dy}{d^2+y^2} = 2\pi\sigma\lambda t.$$

The potential difference between two points distant r_1, r_2 from one wire is $\lambda \log (r_2/r_1)$. Hence the potential difference between the electrodes is

$$2\lambda \log \frac{2d-c}{c} \approx 2\lambda \log \frac{2d}{c}.$$

The resistance

$$R = \frac{V}{I} = \frac{1}{\pi\sigma t} \log \frac{2d}{c}.$$

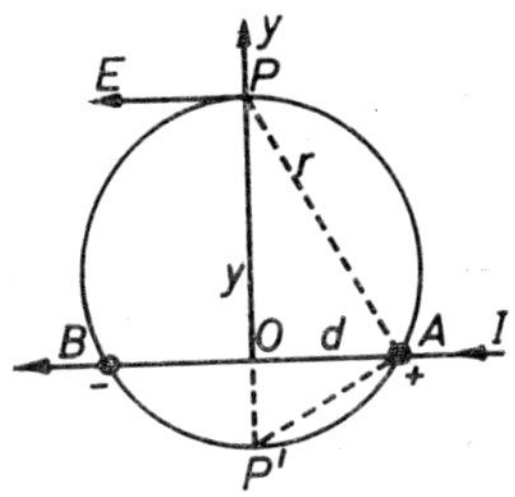

FIG. 25

5.13. *Find the resistance between two small circular electrodes if their centres lie on the circumference of a circular plate.*

We can always cut the infinite plate of the preceding problem along a line of flow extending from one electrode to the other without alteration of the current distribution in the remainder. We can thus choose two lines of flow, as indicated in Fig. 25, which leave the electrodes A and B at the ends of the chord of a circle, the length of the chord being $2d$. The field distribution within this circular section will be as before and the only difference in the calculation will be a change in the limits of integration in the evaluation of I. Here

$$I = 2\lambda\, d\sigma t \int_{P'}^{P} \frac{dy}{d^2+y^2} = 2\lambda\sigma t \left[\tan^{-1}\frac{y}{d}\right]_{P'}^{P}$$

$$= 2\lambda\sigma t \frac{\pi}{2} \quad \text{since} \quad \angle PAP' = \pi/2.$$

Hence

$$R = \frac{2}{\pi\sigma t} \log \frac{2d}{c}.$$

5.14. *Two small circular electrodes, radius c, are situated at a distance 2a apart on the central line of a conducting strip of infinite length, width d, thickness t, conductivity σ. Find the resistance between the electrodes.*

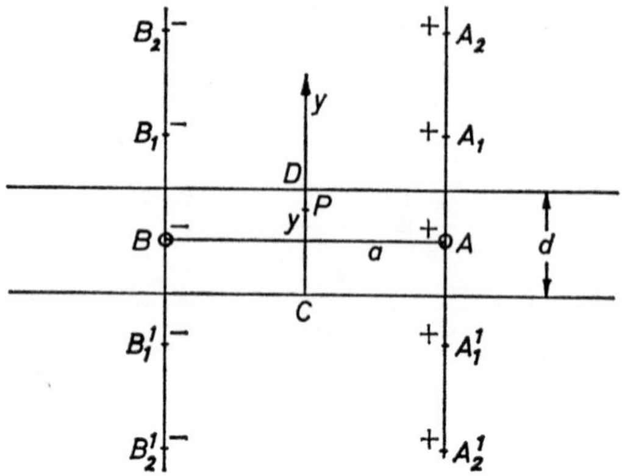

FIG. 26

We can solve this problem by the method of images. The electric field within the strip at any point will be the same as that due to an infinite series of parallel wires, positively charged, spaced d apart on a line through A normal to AB, and a similar infinite set of wires negatively charged on a parallel line through B.

Proceeding by a method similar to the previous problems we have to find the electric field at a point P on the bisector of AB. Taking the field due to a wire as $(\lambda/r)\hat{\mathbf{r}}$ then at P (coord. y) the field due to A and B

$$E = \frac{2\lambda}{r}\cdot\frac{a}{r} = \frac{2\lambda a}{a^2+y^2}.$$

Hence for A and B

$$\int_C^D E\,dy = 2\lambda\int_{-d/2}^{+d/2}\frac{a\,dy}{a^2+y^2}.$$

For A_1 and B_1

$$\int_C^D E\,dy = 2\lambda\int_{-3d/2}^{-d/2}\frac{a\,dy}{a^2+y^2},$$

and for A_1' and B_1'

$$\int_C^D E\,dy = 2\lambda \int_{d/2}^{+3d/2} \frac{a\,dy}{a^2+y^2}.$$

Hence for A, B and all the images

$$\int_C^D E\,dy = 2\lambda \int_{-\infty}^{\infty} \frac{a\,dy}{a^2+y^2} = 2\pi\lambda$$

or the current $I = \sigma t \int_C^D E\,dy = 2\pi\lambda\sigma t.$

The difference of potential between the electrodes due to A, B

$$= 2\lambda \log \frac{2a-c}{c} = 2\lambda \log 2a/c \quad (c \text{ small}),$$

and due to A_1, A_1', B_1 and B_1'

$$4\lambda \log \left[\frac{d^2+(2a-c)^2}{d^2+c^2}\right]^{\frac{1}{2}} = 2\lambda \log \frac{d^2+4a^2}{d^2} = 2\lambda \log \left(1+\frac{4a^2}{d^2}\right).$$

Proceeding in the same way for the other images we obtain

$$V = 2\lambda \left[\log \frac{2a}{c} + \log\left(1+\frac{4a^2}{d^2}\right) + \log\left(1+\frac{4a^2}{4d^2}\right) + \ldots\right],$$

$$= 2\lambda \log \left[\frac{2a}{c}\left(1+\frac{4a^2}{d^2}\right)\left(1+\frac{4a^2}{4d^2}\right)\ldots\right].$$

Hence

$$R = \frac{I}{V} = \frac{1}{\pi\sigma t} \log \left[\frac{2a}{c} \prod_{n=1}^{n=\infty} \left(1+\frac{4a^2}{n^2 d^2}\right)\right] = \frac{1}{\pi\sigma t} \log \left[\frac{d \sinh\left(\dfrac{2\pi a}{d}\right)}{\pi c}\right].$$

$$\left(\sinh x = x \prod_{n=1}^{\infty} \left(1+\frac{x^2}{n^2\pi^2}\right) \text{ from mathematical tables.}\right)$$

EXERCISES

1. Two similar point charges are placed at a distance $2d$ apart. Show that the radius of an earthed conducting sphere which, when placed midway between the charges, neutralises their mutual repulsion, is approximately $d/8$. Estimate the accuracy of this result.

2. A point charge q is placed at a point P inside an earthed conducting spherical shell of radius a at a distance b from the centre O of the sphere. Calculate the potential at any point at a distance z from O in the plane through O perpendicular to OP. Hence calculate the surface charge density on the inside of the spherical shell where this plane intersects it.

3. An earthed conducting sphere of radius r and a ring of thin wire of radius R, $(R > r)$, have their centres coincident. A charge Q is placed on the wire. Calculate the electric field at a point at distance z from the centre of the sphere on a line normal to the plane of the ring and the surface density of charge on the sphere at the point where this line intersects it.

4. Two infinite plane parallel conducting surfaces are separated by a distance d and have potentials of zero and V_0 respectively. A conducting cylinder of infinite length, whose cross-section is a semicircle of radius a, is fixed with its flat face in contact with the earthed plane. Assuming that $d \gg a$, calculate the surface density of charge induced at all points on the earthed plane and the force per unit length acting on the cylinder.

5. A solid metal sphere is divided into two equal halves by a plane through its centre and the two halves are separated by a distance small compared with the radius a of the sphere. The sphere is placed in a uniform electrostatic field produced by distant sources. The direction of the field is normal to the plane of division and its magnitude is E_0. Calculate the magnitude and direction of the force on each half

(*a*) if they are connected by a thin wire and

(*b*) if they are insulated from each other. Assume that both halves are uncharged before the field E_0 is applied.

6. Derive the coefficients of potential in the case of a spherical condenser consisting of an inner sphere of radius a and a concentric outer shell of inside radius b and outside radius c.

The inner and outer conductors of this condenser are connected together and a point charge q is detached from the inner sphere and moved radially at a uniform velocity v to the outer one. Show that the current flowing in the wire connecting the two spheres is given as a function of time t by

$$\frac{qv}{(a+vt)^2} \times \frac{ab}{(b-a)}.$$

7. A small circular circuit of radius a, current i, is placed at a distance d from the plane face of a very thick slab of iron of relative permeability $\bar{\mu}$. Determine the force and the couple on the loop when the plane of the loop is (a) parallel and (b) perpendicular to the surface of the slab.

8. A condenser is designed so that it will not lose through its own leakage resistance more than 5% of its initial charge during a period of one day. Show that this condition specifies a minimum value of the ratio of the dielectric constant to the conductivity and calculate this value.

9. A spherical cavity of radius a exists in an infinite medium of uniform conductivity. At large distances from the cavity the current is uniform and in the direction of the z-axis. Show that the potential is of the form

$$V = A(r + a^3/2r^2) \cos \theta,$$

and that the lines of flow are given by

$$r^3 - a^3 = Cr \operatorname{cosec}^2 \theta,$$

where A and C are constants.

10. A current i flows in a long straight wire parallel to the axis of a circular cylinder of relative permeability $\bar{\mu}$. Show that inside the cylinder the magnetic induction is $2\bar{\mu}/(\bar{\mu}+1)$ times as large as it would be if the cylinder were removed, and is in the same direction.

6 MAGNETIC FIELD OF CURRENTS AND INTERACTION BETWEEN CIRCUITS

The forces which arise between two circuits can be derived from the expression for the force between two current elements $\boldsymbol{dl}_1$, $\boldsymbol{dl}_2$ carrying currents i_1, i_2 (the directions of $\boldsymbol{dl}_1$, $\boldsymbol{dl}_2$ being determined by i_1, i_2). The force on $\boldsymbol{dl}_1$

$$\boldsymbol{dF}_1 = \frac{\mu_0}{4\pi} i_1 i_2 \frac{\boldsymbol{dl}_1 \times (\boldsymbol{dl}_2 \times \boldsymbol{r})}{r^3}$$

where $\boldsymbol{r}$ is drawn from $\boldsymbol{dl}_1$ to $\boldsymbol{dl}_2$.

This force is due to the action of the magnetic field produced by i_2 in $\boldsymbol{dl}_2$ on the current i_1. The magnetic induction due to $\boldsymbol{dl}_2$

$$\boldsymbol{dB}_2 = \frac{\mu_0}{4\pi} i_2 \left(\frac{\boldsymbol{dl}_2 \times \boldsymbol{r}}{r^3}\right).$$

Thus we can write

$$\boldsymbol{dF}_1 = i_1(\boldsymbol{dl}_1 \times \boldsymbol{dB}_2).$$

For a complete circuit, current i, the value of $\boldsymbol{B}$ at any point P, will be

$$\boldsymbol{B} = \frac{\mu_0}{4\pi} \oint \frac{\boldsymbol{dl} \times \boldsymbol{r}}{r^3},$$

where $\boldsymbol{r}$ is drawn from $\boldsymbol{dl}$ to P and the integral is taken round the circuit. The force on a circuit placed in a magnetic field of induction $\boldsymbol{B}$ will be given by $\boldsymbol{F} = i \oint (\boldsymbol{dl} \times \boldsymbol{B})$.

The total force on circuit (1) due to circuit (2)

$$F_1 = \frac{\mu_0}{4\pi} i_1 i_2 \oint_1 \oint_2 \frac{\boldsymbol{dl}_1 \times (\boldsymbol{dl}_2 \times \boldsymbol{r})}{r^3}.$$

Expanding the triple vector product

$$\frac{\boldsymbol{dl}_1 \times (\boldsymbol{dl}_2 \times \boldsymbol{r})}{r^3} = -\boldsymbol{dl}_1 \times \left[\boldsymbol{dl}_2 \times \nabla\left(\frac{1}{r}\right)\right],$$

$$= -\boldsymbol{dl}_2\left[\boldsymbol{dl}_1 \,.\, \nabla\left(\frac{1}{r}\right)\right] + \nabla\left(\frac{1}{r}\right)(\boldsymbol{dl}_1 \,.\, \boldsymbol{dl}_2).$$

But

$$\oint_1 \boldsymbol{dl}_1 \,.\, \nabla\left(\frac{1}{r}\right) = \oint_1 \frac{\partial}{\partial l_1}\left(\frac{1}{r}\right) dl_1 = 0.$$

Hence

$$\boldsymbol{F} = \frac{\mu_0}{4\pi} i_1 i_2 \oint_1 \oint_2 \nabla\left(\frac{1}{r}\right) \boldsymbol{dl}_1 \,.\, \boldsymbol{dl}_2.$$

In Cartesians if $\boldsymbol{dl}_1$ is at point (x, y, z) and $\boldsymbol{dl}_2$ at (x', y', z') and ϵ is the angle between $\boldsymbol{dl}_1$ and $\boldsymbol{dl}_2$

$$F_x = \frac{\mu_0}{4\pi} i_1 i_2 \oint_1 \oint_2 \frac{x'-x}{r^3} \cos \epsilon \, dl_1 \, dl_2.$$

Force on moving charge. Let a particle, charge q, be moving with velocity $\mathbf{v}$ in a field of induction $\boldsymbol{B}$. Since a current flowing in a wire may be regarded as a linear charge in motion we can replace $i\,\boldsymbol{dl}$ in the above expressions by $q\mathbf{v}$. The force on the charge

$$\boldsymbol{F} = q(\mathbf{v} \times \boldsymbol{B}).$$

If we add the force on the charge due to the electric field $\boldsymbol{E}$

$$\boldsymbol{F} = q[\boldsymbol{E} + (\mathbf{v} \times \boldsymbol{B})] \quad \text{(Lorentz force).}$$

The magnetic induction due to a moving charge

$$\boldsymbol{B} = \frac{\mu_0}{4\pi} \frac{q(\mathbf{v} \times \boldsymbol{r})}{r^3}.$$

This holds only for $v \ll c$.

If $\boldsymbol{A}$ = vector potential, $\boldsymbol{B} = \nabla \times \boldsymbol{A}$. For a circuit

$$\boldsymbol{A} = \frac{\mu_0}{4\pi} \oint \frac{i\,\boldsymbol{dl}}{r}.$$

Mutual inductance between two circuits,

$$M = \int_1 \boldsymbol{B} \, . \, d\boldsymbol{S} = \int_1 (\nabla \times \boldsymbol{A}) \, . \, d\boldsymbol{S} = \oint_1 \boldsymbol{A} \, . \, d\boldsymbol{l}_1$$

where $\boldsymbol{B}$, $\boldsymbol{A}$ are due to one ampère in (2).

Hence

$$M = \frac{\mu_0}{4\pi} \oint_1 \oint_2 \frac{d\boldsymbol{l}_1 \, . \, d\boldsymbol{l}_2}{r} = 10^{-7} \oint_1 \oint_2 \frac{d\boldsymbol{l}_1 \, . \, d\boldsymbol{l}_2}{r} \text{ H.}$$

The potential energy of a circuit $W = -i\phi$, where ϕ is the flux through the circuit due to external fields. The force on a circuit can thus be found from $\boldsymbol{F} = -\nabla W = i\nabla\phi$. If we have two circuits, currents i_1 and i_2,

$$W = -i_1 i_2 M \quad \text{and} \quad \boldsymbol{F} = i_1 i_2 \nabla M$$

or

$$F_x = i_1 i_2 \frac{\partial M}{\partial x}.$$

Magnetic dipoles. A small current loop of area $d\boldsymbol{S}$ has a dipole field similar to that of a small magnetic needle. The dipole moment $\boldsymbol{m} = i \, d\boldsymbol{S}$. The potential energy of a dipole in a magnetic field $= -(\boldsymbol{m} \, . \, \boldsymbol{B})$. The torque on the dipole $= \boldsymbol{m} \times \boldsymbol{B}$. The force on the dipole $= -(\boldsymbol{m} \, . \, \nabla)\boldsymbol{B}$. At a point P distant $\boldsymbol{r}$ from the dipole the magnetostatic potential

$$\Omega = -\frac{1}{4\pi}\left[\boldsymbol{m} \, . \, \nabla_P \left(\frac{1}{r}\right)\right] = \frac{1}{4\pi} \frac{(\boldsymbol{m} \, . \, \boldsymbol{r})}{r^3}.$$

At P the vector potential

$$\boldsymbol{A} = -\frac{\mu_0}{4\pi}\left[\boldsymbol{m} \times \nabla_P \left(\frac{1}{r}\right)\right] = \frac{\mu_0}{4\pi} \frac{(\boldsymbol{m} \times \boldsymbol{r})}{r^3}.$$

The potential energy of one dipole in the field of another

$$W = \frac{\mu_0}{4\pi}\left[\frac{(\boldsymbol{m}_1 \, . \, \boldsymbol{m}_2)}{r^3} - \frac{3(\boldsymbol{m}_1 \, . \, \boldsymbol{r})(\boldsymbol{m}_2 \, . \, \boldsymbol{r})}{r^5}\right],$$

$$= \frac{\mu_0}{4\pi} \frac{m_1 m_2}{r^3} (\cos \epsilon - 3 \cos \theta_1 \cos \theta_2)$$

where $\boldsymbol{r}$ is the vector distance between them, ϵ is the angle between $\boldsymbol{m}_1$ and $\boldsymbol{m}_2$ and θ_1, θ_2 are the angles between $\boldsymbol{m}_1$ and $\boldsymbol{r}_1$, $\boldsymbol{m}_2$ and $\boldsymbol{r}_2$.

6.1. *Two circular loops of radii a and b carrying currents i_1 and i_2 respectively are mounted coaxially. Show that, if b is small compared with a, the force between the loops is a maximum when the distance between them is a/2 and find the magnitude of this force.*

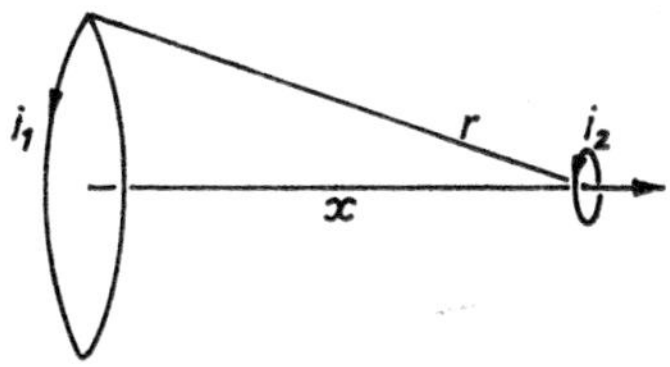

FIG. 27

Let the coils be distance x apart. We can assume the magnetic field is uniform over the area of the small circle.

$$B = \frac{\mu_0 i_1}{2}\frac{a^2}{r^3}.$$

The flux through the small loop $= \pi b^2 . (\mu_0 i_1/2)(a^2/r^3) = \phi$. The potential energy,

$$W = -i_2\phi = -\mu_0 \pi a^2 b^2 i_1 i_2 \frac{1}{2(x^2+a^2)^{\frac{3}{2}}}.$$

The force between them $= -\partial W/\partial x$.

$$|F| = \tfrac{1}{2}\mu_0 \pi a^2 b^2 i_1 i_2 (3x/(x^2+a^2)^{\frac{5}{2}}).$$

F will be attractive if i_1 and i_2 are in the same direction of rotation and repulsive if in opposite directions.

F will have a turning value when

$$\frac{d}{dx}\left[\frac{x}{(x^2+a^2)^{\frac{5}{2}}}\right] = 0, \quad (x^2+a^2)^{-\frac{5}{2}} - 5x(x^2+a^2)^{-\frac{7}{2}}\,x = 0,$$

i.e.

$$x = a/2.$$

This clearly gives a maximum since the force is zero for $x \to 0$ or $x \to \infty$.

Maximum

$$F = \frac{\mu_0 i_1 i_2 \pi b^2}{a^2}\cdot\frac{24}{25\sqrt{5}}.$$

Otherwise, we may treat the small loop as a dipole of moment $m = \pi b^2 i_2$ in the x-direction. Since the force on a dipole is $(\boldsymbol{m} \,.\, \nabla)\boldsymbol{B}$ we have

$$F_x = m\frac{\partial}{\partial x}B_x = (\pi b^2 i_2)\frac{\partial}{\partial x}\left[\frac{\mu_0 i_1}{2}\frac{a^2}{(a^2+x^2)^{\frac{3}{2}}}\right] = F \quad \text{above.}$$

6.2. *A long straight wire lies in the plane of a circle of radius a the centre of which is at a perpendicular distance d from the wire. Find the mutual inductance.*

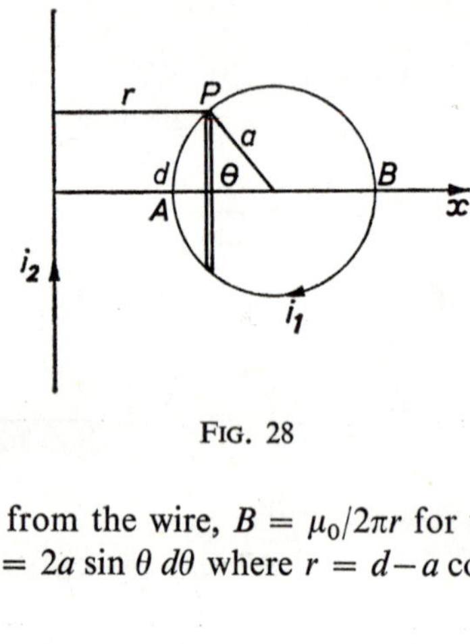

FIG. 28

At P, distant r from the wire, $B = \mu_0/2\pi r$ for unit current. Area of strip width $dr = 2a \sin\theta \, d\theta$ where $r = d - a\cos\theta$. Flux through circle $= M$.

$$M = \frac{\mu_0}{2\pi}\int_A^B \frac{2a\sin\theta \, dr}{r} = \frac{\mu_0}{\pi}\int_0^\pi \frac{a^2\sin^2\theta}{d-a\cos\theta}\,d\theta,$$

$$= \frac{\mu_0}{\pi}\int_0^\pi \frac{(d^2-a^2\cos^2\theta)-(d^2-a^2)}{d-a\cos\theta}\,d\theta,$$

$$= \frac{\mu_0}{\pi}\left\{\int_0^\pi (d+a\cos\theta)\,d\theta-(d^2-a^2)\int_0^\pi \frac{d\theta}{d-a\cos\theta}\right\},$$

$$= \frac{\mu_0}{\pi}\left\{d\pi-(d^2-a^2)\frac{\pi}{\sqrt{(d^2-a^2)}}\right\} = \mu_0\{d-\sqrt{(d^2-a^2)}\},$$

$$= 4\pi 10^{-7}\{d-\sqrt{(d^2-a^2)}\}.$$

6.3. *Find the force between the wire and the circle for currents i_1, i_2 in Fig. 28.*

The force will be perpendicular to the wire of magnitude

$$F_x = i_1 i_2 \frac{\partial M}{\partial x} = \mu_0 i_1 i_2 \frac{\partial}{\partial x}\{x-\sqrt{x^2-a^2}\}, \quad (x = d)$$

$$= -\mu_0 i_1 i_2 \{d/\sqrt{(d^2-a^2)}-1\} \text{ N.}$$

For the directions of i_1, i_2 chosen in the figure the force is attractive.

6.4. *Find the mutual inductance between two square circuits which bound the opposite faces of a cube.*

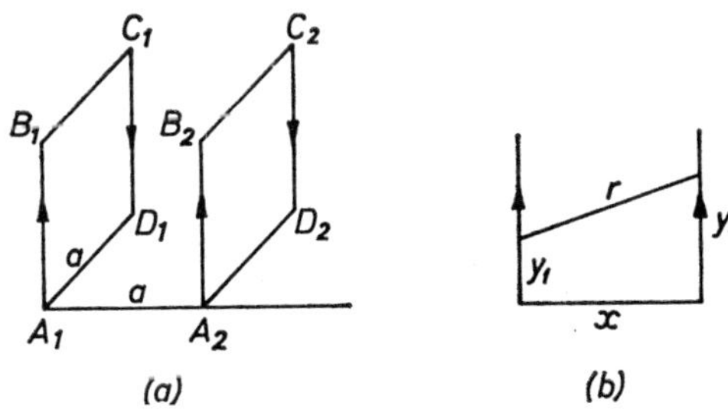

FIG. 29

We have

$$M = \frac{\mu_0}{4\pi} \oint_1 \oint_2 \frac{\cos \epsilon}{r}\, dl_1\, dl_2.$$

Take dl_1 on A_1B_1. If dl_2 is on A_2B_2, $\cos \epsilon = 1$, and if dl_2 is on C_2D_2, $\cos \epsilon = -1$. For the other sides B_2C_2, D_2A_2, $\cos \epsilon = 0$.

Since, by symmetry, the contribution of each side of circuit (1) to the integral is the same, we can write

$$M = 4\frac{\mu_0}{4\pi}\left[\int_{A_1}^{B_1}\int_{A_2}^{B_2}\frac{dl_1\, dl_2}{r} - \int_{A_1}^{B_1}\int_{C_2}^{D_2}\frac{dl_1\, dl_2}{r}\right].$$

If we perform the integration for two wires of length *a*, distance *x* apart we can apply the result to find both integrals. (Fig. 29(b).)

$$\int_0^a dy_1 \int_0^a \frac{dy_2}{\sqrt{[(y_2-y_1)^2+x^2]}} = \int_0^a dy_1 \left[\sinh^{-1}\frac{y_2-y_1}{x}\right]_0^a,$$

$$= \int_0^a dy_1 \left[\sinh^{-1}\frac{a-y_1}{x}+\sinh^{-1}\frac{y_1}{x}\right]$$

$$= 2\int_0^a \left(\sinh^{-1}\frac{y_1}{b}\right) dy_1,$$

$$= 2\left[y_1 \sinh^{-1}\frac{y_1}{x}\right]_0^a - 2\int_0^a \frac{y_1\, dy_1}{\sqrt{x^2+y_1^2}},$$

$$= 2\left[a \sinh^{-1}\frac{a}{x} - \sqrt{(a^2+x^2)}+x\right],$$

$$= 2\left[a \log\frac{a+\sqrt{a^2+x^2}}{x} - \sqrt{(a^2+x^2)}+x\right].$$

If we put $x = a$, $x = \sqrt{2}a$ in turn, we find

$$M = 8\frac{\mu_0}{4\pi}\left\{[a\log(1+\sqrt{2})-a\sqrt{(2)}+a] - \left[a\log\frac{1+\sqrt{3}}{\sqrt{2}}-\sqrt{3}a+\sqrt{2}a\right]\right\},$$

$$= 8a\left[\log\frac{2+\sqrt{2}}{1+\sqrt{3}}+1-2\sqrt{2}+\sqrt{3}\right]10^{-7} \text{ H.}$$

6.5. *A circuit in the form of a circle of radius a, carrying a current* i_1*, is placed with its centre at a distance* $b(b > a)$ *from a long straight wire carrying a current* i_2*, the wire being at right angles to the plane of the circle. Show that the couple on the circle about a diameter which passes through the straight wire is* $\mu_0 i_1 i_2 a^2/2b$*.*

Show that if $b < a$ *the couple is* $\mu_0 i_1 i_2 b/2$*.*

Consider the element $\boldsymbol{dl}$ at point P. Then $dl = a\,d\theta$ and $B = \mu_0 i_2/2\pi r$ in the direction at right angles to AP. The force on the element is i_1 $(\boldsymbol{dl} \times \boldsymbol{B})$, which is normal to the plane of the circle and of magnitude

$$\frac{\mu_0 i_1 i_2 \sin\psi a\,d\theta}{2\pi r}.$$

The total couple on the circle about AC is thus

$$\frac{2\mu_0 i_1 i_2 a^2}{2\pi}\int_0^\pi \frac{\sin\theta \sin\psi}{r}\,d\theta.$$

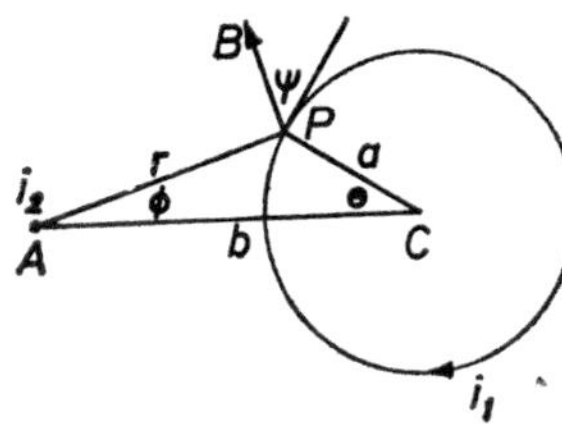

FIG. 30

Since $\sin\psi/b = \sin\theta/r$, we have

$$\text{couple} = \frac{\mu_0 i_1 i_2 a^2 b}{\pi}\int_0^\pi \frac{\sin^2\theta}{r^2}\,d\theta.$$

$$\int_0^\pi \frac{\sin^2\theta}{r^2}\,d\theta = \int_0^\pi \frac{\sin^2\theta}{b^2+a^2-2ab\cos\theta}\,d\theta$$

$$= \frac{1}{2ab}\int_0^\pi \frac{(1-\cos^2\theta)\,d\theta}{\alpha-\cos\theta} \quad \text{where} \quad \alpha = \frac{b^2+a^2}{2ab},$$

$$= \frac{1}{2ab}\int_0^\pi \frac{(\alpha^2-\cos^2\theta)-(\alpha^2-1)}{\alpha-\cos\theta}\,d\theta,$$

$$= \frac{1}{2ab}\int_0^\pi \left[(\alpha+\cos\theta)-\frac{(\alpha^2-1)}{\alpha-\cos\theta}\right]d\theta,$$

$$= \frac{1}{2ab}\left[\alpha\pi-(\alpha^2-1)\frac{\pi}{\sqrt{\alpha^2-1}}\right] = \frac{\pi}{2ab}(\alpha-\sqrt{\alpha^2-1}),$$

$$= \frac{\pi}{2ab}\left[\frac{b^2+a^2}{2ab}-\frac{b^2-a^2}{2ab}\right] = \frac{\pi}{2b^2}.$$

Hence the couple on the loop $= \mu_0 i_1 i_2 a^2/2b$. If $b < a$, i.e. the straight wire lies within the circle, then $\sqrt{\alpha^2-1} = (a^2-b^2)/2ab$ and the couple would be $\mu_0 i_1 i_2 b/2$.

Note that in this problem the torque on the circle for $b > a$ is the same if the circle is replaced by a dipole of moment $\pi a^2 i_1$. This is not usually permissible. It is merely a result of the geometry of this particular problem. Let us solve this problem by replacing the circuit by its equivalent magnetic shell of dipole moment i_1 per unit area.

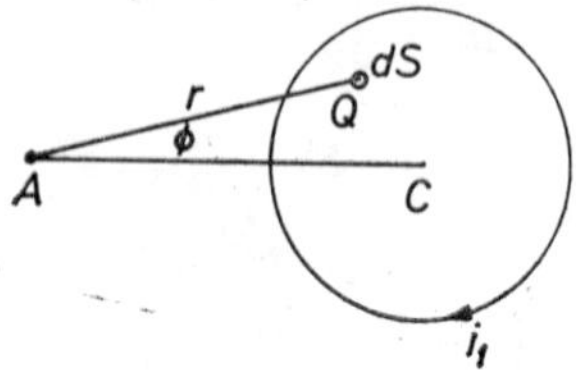

FIG. 31

Consider small area $\boldsymbol{dS}$ at Q. The couple on the dipole of moment $i_1\, \boldsymbol{dS}$ is $(\mu_0/2\pi r) i_1 i_2\, dS$ about AQ. Hence about AC

$$\text{couple} = \frac{\mu_0 i_1 i_2}{2\pi} \int \frac{\cos\phi}{r}\, dS \quad \text{over the circular area.}$$

The integral is the same as that which would appear in the two-dimensional problem of finding the magnetic field due to a current uniformly distributed over the cross-section of a solid cylinder. We know in that case that the field is the same as if the whole current I flowed in a thin wire through C. If j is the current density the magnetic field at A would be

$$H = \int \frac{j\, dS \cos\phi}{2\pi r} = \frac{\pi a^2 j}{2\pi b},$$

since $H = I/2\pi b$ where $I = \pi a^2 j$.

Hence

$$\int \frac{\cos\phi}{r}\, dS = \frac{\pi a^2}{b} \quad \text{for } A \text{ outside.}$$

If the point A is inside the circle

$$H = \frac{\pi b^2 j}{2\pi b} = \frac{b}{2} j$$

and

$$\int \frac{\cos\phi}{r}\, dS = \pi b$$

Thus the torque for $b > a$ is $\mu_0 i_1 i_2 a^2/2b$ and for $b < a$, $\mu_0 i_1 i_2 b/2$.

6.6. *A long cylindrical solenoid of radius r and n turns per unit length carries a current i. Show that, if end effects are neglected,*

(*a*) *the force which tends to decrease the length of the solenoid is*

$$\tfrac{1}{2}\mu_0 n^2 i^2 \pi r^2;$$

(*b*) *the force per unit area which tends to increase the radius of the coil is*

$$\tfrac{1}{2}\mu_0 n^2 i^2.$$

(*a*) The force between two current elements in the same direction is attractive. Hence a solenoid will tend to diminish in length. The energy associated with the magnetic field $= \tfrac{1}{2}Li^2$ where L is the self-inductance. Neglecting end effects

$$L = \mu_0 n^2 \pi r^2 l,$$

where l = length of coil. The force tending to contract the coil will be given by $F = -\tfrac{1}{2}i^2(\partial L/\partial l)$. Since a change of l alters n we substitute $N = nl$, the total number of turns, which is constant.

$$L = \mu_0 N^2 \pi r^2 / l,$$

so that

$$F = \tfrac{1}{2}i^2 \mu_0 N^2 \pi r^2 \left(\frac{1}{l^2}\right) = \tfrac{1}{2}\mu_0 n^2 i^2 \pi r^2.$$

(*b*) If F' is the force per unit area radially outwards the total work done when r increases to $r+\delta r$ is $2\pi r l F'\,\delta r$. The increase in energy $= \tfrac{1}{2}i^2(\partial L/\partial r)\,\delta r$.

Hence

$$F' = \frac{i^2}{4\pi l r}\frac{\partial L}{\partial r} = \frac{i^2}{4\pi l r}(2\mu_0 n^2 \pi r l) = \tfrac{1}{2}\mu_0 n^2 i^2.$$

Note that in both cases the forces act in such a direction as to increase the self-inductance of the coil.

6.7. *Two small magnets are mounted so that one lies along the line joining their centres, and the other lies at right angles to it. Prove that the action of each magnet on the other reduces to a single force at right angles to the line joining their centres, acting at a point on that line one third of its length from the longitudinal magnet.*

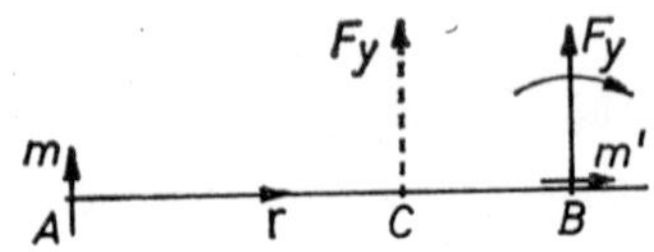

FIG. 32

Potential energy

$$W = \frac{\mu_0}{4\pi}\frac{mm'}{r^3}(\cos\epsilon - 3\cos\theta\cos\theta').$$

For coplanar dipoles

$$W = \frac{\mu_0}{4\pi}\frac{mm'}{r^3}(\sin\theta\sin\theta' - 2\cos\theta\cos\theta'),$$

since $\epsilon = \theta' - \theta$. Here $\theta = \pi/2$, $\theta' = 0$, $\epsilon = \pi/2$. The couple on

$$m' = -\frac{\partial W}{\partial\theta'} = -\frac{\mu_0}{4\pi}\frac{mm'}{r^3}(\sin\theta\cos\theta' + 2\cos\theta\sin\theta') = -\frac{\mu_0}{4\pi}\frac{mm'}{r^3}.$$

The force on a dipole in a variable field is given by

$$\boldsymbol{F} = (\boldsymbol{m}\,.\,\nabla)\boldsymbol{B}.$$

If we take the y-axis normal to $\boldsymbol{r}$, then

$$F_y = \left(m'\frac{\partial}{\partial r}\right)B_y = m'\frac{\partial}{\partial r}\left(-\frac{\mu_0}{4\pi}\frac{m}{r^3}\right) = \frac{\mu_0}{4\pi}\frac{3mm'}{r^4}.$$

The action of m on m' thus consists of the above couple and force acting in the directions shown in Fig. 32. These can be replaced by the single force F_y acting at the point C distant $r/3$ from B.

The student should verify that the couple on m due to m' is

$$-\frac{\mu_0}{4\pi}\cdot\frac{2mm'}{r^3}$$

and that the force perpendicular to AB is

$$-\frac{\mu_0}{4\pi}\cdot\frac{3mm'}{r^4}$$

Hence the action of m' on m may be represented by a force through C equal and opposite to F_y.

6.8. *Two dipoles each of moment m are placed in a uniform magnetic field of induction* **B** *which is at right angles to the line joining their centres. Find the positions of equilibrium*

(*a*) *if the dipoles are constrained to rotate in the plane through their centres parallel to* **B**, *and*

(*b*) *if they are constrained to rotate in planes normal to the line joining their centres.*

(*a*) Let the dipoles be inclined to $\boldsymbol{B}$ at angles θ, ϕ. Then the total potential energy of the system is

$$W = \frac{\mu_0}{4\pi}\frac{m^2}{r^3}[\cos(\phi-\theta)-3\sin\theta\sin\phi]-mB(\cos\theta+\cos\phi).$$

For equilibrium

$$\frac{\partial W}{\partial\theta} = 0 \quad\text{and}\quad \frac{\partial W}{\partial\phi} = 0,$$

which gives

$$\frac{\mu_0}{4\pi}\frac{m^2}{r^3}[-\sin\theta\cos\phi-2\cos\theta\sin\phi]+mB\sin\theta = 0,$$

$$\frac{\mu_0}{4\pi}\frac{m^2}{r^3}[-\cos\theta\sin\phi-2\sin\theta\cos\phi]+mB\sin\phi = 0,$$

or

$$\sin\theta\cos\phi+2\cos\theta\sin\phi = \frac{4\pi r^3 B}{\mu_0 m}\sin\theta,$$

$$\cos\theta\sin\phi+2\sin\theta\cos\phi = \frac{4\pi r^3 B}{\mu_0 m}\sin\phi.$$

By inspection $\theta = \phi$ and $\theta = -\phi$ are the solutions.

$$\theta = \phi \quad \text{gives} \quad \theta = 0, \phi = 0 \quad \text{or} \quad \cos\phi = \frac{4\pi r^3 B}{3\mu_0 m},$$

$$\theta = -\phi \quad \text{gives} \quad \cos\phi = -\frac{4\pi r^3 B}{\mu_0 m}.$$

(*b*) In this case the potential energy is

$$W = \frac{\mu_0 m^2}{4\pi r^3}[\cos(\phi-\theta)]-mB(\cos\theta+\cos\phi),$$

since the dipoles are at right angles to *r*.

$$\frac{\partial W}{\partial\theta} = 0, \quad \frac{\partial W}{\partial\phi} = 0$$

gives

$$\sin(\phi-\theta) = -\frac{4\pi r^3 B}{\mu_0 m}\sin\theta,$$

$$-\sin(\phi-\theta) = -\frac{4\pi r^3 B}{\mu_0 m}\sin\phi.$$

Solutions: $\theta = 0, \phi = 0,$

or

$$\theta = -\phi, \quad \cos\phi = \frac{4\pi r^3 B}{2\mu_0 m}.$$

The student should investigate the conditions of stability in the above positions.

6.9. *A small sphere of radius a and relative permeability $\bar{\mu}$ is placed near a circuit which, when carrying unit current, would produce a magnetic field H at the centre of the sphere. Show that the presence of the sphere increases the self-inductance of the circuit by approximately* $4\pi a^3 \mu_0(\bar{\mu}-1)H^2/(\bar{\mu}+2)$.

Let us first find the flux ϕ through the circuit produced by a dipole $\boldsymbol{m}$ placed near to it. If the current is i then the potential energy of the circuit in the field of the dipole $= -i\phi$. This must be the same as the potential energy of the dipole in the field of the circuit which is $-\boldsymbol{m} \cdot \boldsymbol{B}$ where $\boldsymbol{B} = -\mu_0 \nabla\Omega = -(\mu_0/4\pi)i\nabla\omega$, ω being the solid angle subtended at $\boldsymbol{m}$ by the circuit.

Hence

$$\phi = -\frac{\mu}{4\pi}\boldsymbol{m} \cdot \nabla\omega.$$

The small sphere placed in a field $\boldsymbol{H}$, assumed uniform in the region of the sphere, will become magnetised uniformly and will produce an external field as if the sphere were replaced by a dipole of moment

$$\boldsymbol{m} = 4\pi a^3 \frac{\bar{\mu}-1}{\bar{\mu}+2}\boldsymbol{H}.$$

The flux through the circuit for unit current will thus be increased by the flux due to this dipole. But at the sphere

$$\boldsymbol{H} = -\frac{1}{4\pi}\nabla\omega \quad \text{or} \quad \nabla\omega = -4\pi\boldsymbol{H}.$$

Hence the self inductance will be increased by

$$\phi = 4\pi a^3 \mu_0(\bar{\mu}-1)H^2/(\bar{\mu}+2).$$

6.10. *A dipole of moment* $\boldsymbol{m}$ *is placed at the centre of a sphere of radius R. Show that the mean value of* $\boldsymbol{B}$ *throughout the volume of the sphere is* $\mu_0\boldsymbol{m}/(2\pi R^3)$.

We shall make use of the vector identity

$$\int (\nabla \times \boldsymbol{A})\, d\tau = \int (\boldsymbol{dS} \times \boldsymbol{A}).$$

Let $\boldsymbol{C}$ be any arbitrary constant vector. Then

$$\nabla \cdot (\boldsymbol{A} \times \boldsymbol{C}) = \boldsymbol{C} \cdot (\nabla \times \boldsymbol{A}) - \boldsymbol{A} \cdot (\nabla \times \boldsymbol{C}) = \boldsymbol{C} \cdot (\nabla \times \boldsymbol{A}).$$

For any volume

$$C . \int (\nabla \times A)\, d\tau = \int \nabla . (A \times C)\, d\tau,$$
$$= \int (A \times C) . dS = C . \int (dS \times A).$$

Since C can have any direction the above result follows.

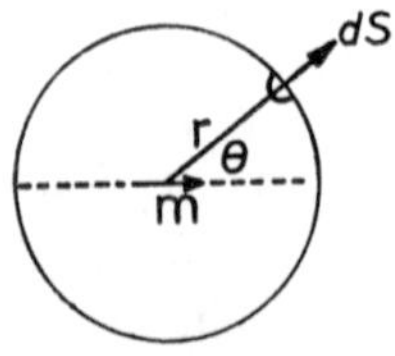

FIG. 33

Let A be the vector potential due to the dipole.

$$A = \frac{\mu_0}{4\pi} \frac{m \times \hat{r}}{r^2} \quad \text{and} \quad B = \nabla \times A.$$

Over the spherical volume

$$\int B\, d\tau = \int (\nabla \times A)\, d\tau = \frac{\mu_0}{4\pi} \int dS \times \left(\frac{m \times \hat{r}}{R^2} \right),$$
$$= \frac{\mu_0}{4\pi} \left[\int m \left(\frac{dS . \hat{r}}{R^2} \right) - \int \frac{\hat{r}}{R^2} (m . dS) \right].$$

The mean value of B by symmetry is in the direction of m, and dS is in the same direction as $\hat{r}$.

Hence

$$\int B\, d\tau = \frac{\mu_0}{4\pi} \left[\int \frac{m\, dS}{R^2} - \int_0^\pi (\cos\theta) \left(\frac{m \cos\theta}{R^2} \right) (2\pi R^2 \sin\theta\, d\theta) \right],$$
$$= \frac{\mu_0}{4\pi} \left[4\pi m - 2\pi m \int_0^\pi \cos^2\theta \sin\theta\, d\theta \right],$$
$$= \frac{\mu_0}{4\pi} \left[4\pi m - 2\pi m \left(\frac{2}{3} \right) \right] = \frac{\mu_0 2m}{3}.$$

Thus the mean value of $B = \frac{2}{3}\mu_0 m / (\frac{4}{3}\pi R^3) = \mu_0 m / (2\pi R^3)$.

Show that the same result is obtained if the dipole is situated at any point within the sphere.

6.11. *A spherical shell of radius a which carries a charge q rotates about a diameter with angular velocity ω. Assuming that the surface distribution of charge remains uniform show that the magnetic field inside the sphere is constant and that outside the sphere the field is the same as that of a magnetic dipole at the centre.*

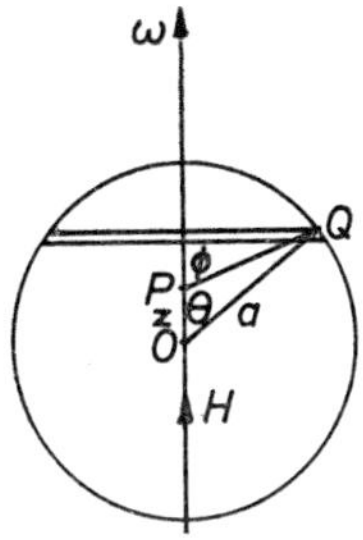

FIG. 34

Let us calculate the magnetic potential at the point P on the axis of rotation distant z from the centre. Take a circular strip formed by rotation of $a\,d\theta$.

Charge on strip $= 2\pi a^2 \sin\theta\, d\theta\sigma$ where σ, the surface density, is $q/4\pi a^2$.

The current produced by the rotation is the charge which flows past a point in unit time and is given by

$$i = 2\pi a^2 \sin\theta\, d\theta\sigma\omega/2\pi = \sigma\omega a^2 \sin\theta\, d\theta.$$

The potential at P due to this current is $i/4\pi$ (solid angle subtended at P) $= (i/4\pi)\,2\pi(1+\cos\phi) = \frac{1}{2}\sigma\omega a^2(1+\cos\phi)\sin\theta\, d\theta$, taking account of the direction of the current.

Hence for whole shell

$$\Omega = \frac{\sigma\omega a^2}{2}\int_0^\pi (1+\cos\phi)\sin\theta\, d\theta,$$

$$= \sigma\omega a^2 + \frac{\sigma\omega a^2}{2}\int_0^\pi \cos\phi\sin\theta\, d\theta,$$

$$\cos\phi = \frac{a\cos\theta - z}{\rho},$$

where

$$\rho = PQ = (a^2+z^2-2az\cos\theta)^{\frac{1}{2}}.$$

Thus

$$\int_0^\pi \cos\phi \sin\theta \, d\theta = \int_0^\pi \frac{(a\cos\theta - z)\sin\theta \, d\theta}{(a^2+z^2-2az\cos\theta)^{\frac{1}{2}}}.$$

Let $u = \cos\theta$, $\alpha = (a^2+z^2)/2az$.

$$\int_0^\pi \cos\phi \sin\theta \, d\theta = \frac{1}{\sqrt{2az}}\int_{-1}^{1} \frac{(au-z)}{\sqrt{\alpha-u}}\, du,$$

$$= \frac{1}{\sqrt{2az}}\{(a\alpha-z)\textstyle\int_{-1}^{1}(\alpha-u)^{-\frac{1}{2}}\,du - a\int_{-1}^{1}(\alpha-u)^{\frac{1}{2}}\,du\},$$

$$= \frac{-2}{\sqrt{2az}}\left\{(a\alpha-z)(\sqrt{\alpha-1}-\sqrt{\alpha+1}) - \frac{a}{3}[(\alpha-1)^{\frac{3}{2}}-(\alpha+1)^{\frac{3}{2}}]\right\}.$$

Inside sphere,

$$z < a \quad \sqrt{\alpha-1} = \frac{a-z}{\sqrt{2az}}, \quad \sqrt{\alpha+1} = \frac{a+z}{\sqrt{2az}}, \quad a\alpha - z = \frac{a^2-z^2}{2z}.$$

Hence

$$\textstyle\int_0^\pi \cos\phi \sin\theta \, d\theta = -4z/3a, \quad \text{and} \quad \Omega_1 = \sigma\omega a^2 - \frac{2}{3}\sigma\omega az. \qquad (1)$$

Outside sphere

$$z > a, \quad \sqrt{\alpha-1} = \frac{z-a}{\sqrt{2az}},$$

and one finds

$$\int_0^\pi \cos\phi \sin\theta \, d\theta = +\frac{2}{3}\frac{a^2}{z^2} - 2$$

$$\Omega_2 = \tfrac{1}{3}\omega\sigma a^4/z^2. \qquad (2)$$

Since the problem has axial symmetry the potential at any point within the sphere can be expressed in terms of Legendre functions

$$\Omega_1 = \sum A_n r^n P_n(\cos\theta).$$

For $\theta = 0$, by comparison with (*1*), we find

$$A_0 = \sigma\omega a^2, \quad A_1 = -\tfrac{2}{3}\sigma\omega a, \qquad {}_2 = A_3 = \ldots = 0.$$

Hence

$$\Omega_1 = \sigma\omega a^2 - \tfrac{2}{3}\sigma\omega ar \text{ c} \qquad\qquad -\tfrac{2}{3}\sigma\omega az.$$

$$H = -\frac{\partial\Omega}{\partial z} = \tfrac{2}{3}\sigma\omega a = \frac{\omega q}{6\pi a}$$

at all points inside. For any point outside we can write

$$\Omega_2 = \sum B_n r^{-(n+1)} P_n(\cos\theta),$$

and by comparison with (*2*) for $\theta = 0$, $B_1 = \frac{1}{3}\sigma\omega a^4$, $B_2 = B_3 \cdots = 0$.

Hence in general

$$\Omega_2 = (\tfrac{1}{3}\sigma\omega a^4)\frac{\cos\theta}{r^2}.$$

This potential is that of a dipole field with a dipole of magnitude $\omega q a^2/3$ at the centre of the sphere in the direction of the axis of rotation.

The problem of the rotating charged shell may also be solved by making use of an extension of Ampère's principle of the equivalent magnetic shell. The magnetic field at any point due to a circuit is the same as that due to any thin shell bounded by the circuit of magnetic moment per unit area equal to the current. If we replace the circular strip by a shell then the magnetic moment per unit area $= \sigma\omega a^2 \sin\theta \, d\theta$. The shell has a thickness $a \sin\theta \, d\theta$ so that we can think of the shell as having a moment per unit volume of $\sigma\omega a$. This is the same for any value of θ. Hence the rotating shell is magnetically equivalent to a uniformly magnetised sphere of intensity of magnetisation $I = \sigma\omega a$. The field outside the sphere will thus be that of a dipole of magnitude $\frac{4}{3}\pi a^3 I = \frac{4}{3}\pi a^4 \sigma\omega = \omega q a^2/3$. For a point inside the sphere the magnetic induction (not H) will be the same in the two cases. For a uniformly magnetised sphere, B is constant and is given by $\frac{2}{3}\mu_0 I = \mu_0 \omega q/6\pi a$. Thus for the rotating charged sphere $H = \omega q/6\pi a$.

(In the case of a real uniformly magnetised sphere we have $H = -I/3$ because at any point we must have $\boldsymbol{B} = \mu_0(\boldsymbol{H}+\boldsymbol{I})$.)

We can verify the above value of H by direct calculation of its value at the centre. Due to strip

$$dH = \frac{i}{2}\frac{a^2 \sin^2 \theta}{a^3} = \frac{i}{2a} \sin^2 \theta$$

$$H = \frac{\sigma\omega a}{2}\int_0^\pi \sin^3 \theta \, d\theta = \tfrac{2}{3}\sigma\omega a.$$

6.12. *A charge q is distributed uniformly over a spherical shell of mass m. A uniform magnetic field of induction $\mathbf{B}$ is established. Show that, if the charge remains uniformly distributed and the shell can rotate without friction, it will acquire an angular velocity $-\mathbf{B}q/2m$.*

What angular velocity would be acquired if the sphere were solid and the charge distributed throughout the sphere with uniform density?

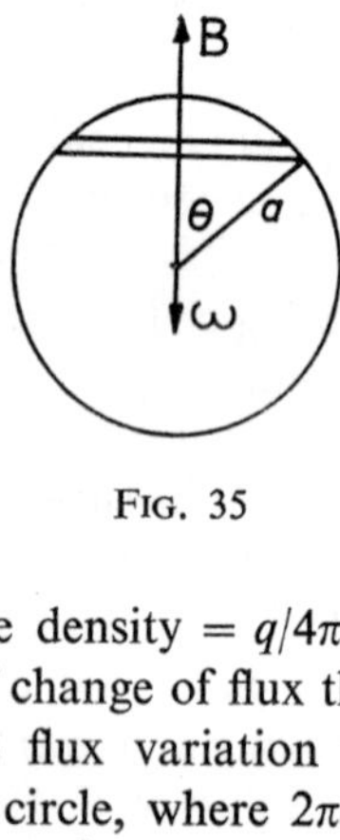

FIG. 35

Let σ = surface charge density = $q/4\pi a^2$. Charge on strip $a\,d\theta$ is $2\pi a^2 \sin\theta\sigma\,d\theta$. Rate of change of flux through circle bounded by strip = $\pi a^2 \sin^2 \theta\dot{B}$. This flux variation will produce an electric field E tangential to the circle, where $2\pi a \sin\theta E = -\pi a^2 \sin^2\theta\dot{B}$. Therefore $E = -(a/2)\sin\theta\dot{B}$. The torque on the strip about a diameter parallel to $\mathbf{B}$ due to E is

$$E 2\pi a^3 \sin^2\theta\sigma\,d\theta = -\pi a^4\sigma\dot{B}\sin^3\theta\,d\theta.$$

$$\text{Hence total torque on sphere} = -\pi a^4\sigma\dot{B}\int_0^\pi \sin^3\theta\,d\theta,$$

$$= -\tfrac{4}{3}\pi a^4\sigma\dot{B} = -\frac{a^2}{3}q\dot{B}.$$

Since the shell is free to rotate, this torque will at any instant be equal to the rate of change of angular momentum. If ω is the angular velocity

$$(\tfrac{2}{3}ma^2)\dot{\omega} = -\frac{a^2}{3}q\dot{B}.$$

Integrating from zero time to the time of establishment of B

$$\omega = -\frac{qB}{2m}.$$

Solid sphere: Let ρ = charge density.

Torque on shell thickness

$$dr = -\frac{r^2}{3}(4\pi r^2 \rho \, dr)\dot{B}$$

$$= -\tfrac{4}{3}\pi\rho\dot{B}r^4 \, dr.$$

Total torque on sphere

$$= -\tfrac{4}{3}\pi\rho\dot{B}\frac{a^5}{5} = -q\frac{a^2}{5}\dot{B}.$$

Rate of change of angular momentum

$$= (\tfrac{2}{5}ma^2)\dot{\omega} = -q\frac{a^2}{5}\dot{B}.$$

Hence $\omega = -qB/2m$, as before.

EXERCISES

1. A current i flows in a wire of length l which forms the periphery of an ellipse of area A. Show that the magnetic field at the centre is $il/4A$.

2. Find an expression for the magnetic field at any point on the axis of a uniformly wound solenoid of finite length.

The following apparatus is employed in a recent method of standardising the ampère. A long solenoid of radius a and n turns m^{-1} is mounted with its axis vertical; the winding is broken at the centre and current is supplied in opposite directions in the two halves. A small flat circular coil of N turns of radius b is suspended from the arm of a balance and hangs inside the solenoid in the central plane with axes coincident. The same current i flows through the two halves of the solenoid and the small coil. Explain why a force on the small coil is produced and show that its magnitude is approximately

$$\mu_0 \pi b^2 n N i^2 / a.$$

3. $ABCD$ is a square circuit of side $2a$, current i_1; E, F, are the mid-points of AD and BC. A long straight wire carrying current i_2, at right angles to the plane of the square, passes through a point O on FE produced, the distance of O from the centre of the square being d. If the angles AOE and BOE are ϕ_1 and ϕ_2 respectively, show that the couple exerted on the square about EF is

$$\frac{\mu_0 i_1 i_2}{\pi}\left[a \log \frac{\sin \phi_1}{\sin \phi_2} + a(\phi_1+\phi_2) - d(\phi_1-\phi_2)\right].$$

4. A circular circuit carrying a current i from west to east is coincident with a parallel of south latitude λ on the earth's surface. Assuming that the earth is a uniformly magnetised sphere of radius a and that the magnetic field on the equator is H, show that the circuit would experience a force towards the south pole of amount

$$6\pi\mu_0 a i \sin \lambda \cos^2 \lambda H.$$

(Neglect any effect of the current on the earth's magnetisation.)

5. Two circular loops of wire of radii a and b lie in the same plane at a distance r apart, a and b being small compared with r. Find the mutual inductance between the loops.

If the loops are rotated about axes, in the initial plane, through their centres and normal to the line joining them, their planes remaining parallel, show that at a certain inclination their mutual inductance is zero.

6. Show that the force between two square circuits which bound the opposite faces of a cube and carry currents i_1, i_2 ampères is

$$4i_1i_2[3\sqrt{2}-2-\sqrt{3}]10^{-7}\ \text{N}.$$

Calculate the force by three methods.

(*a*) Use the result that the force between two parallel wires of length a, distance x apart is

$$\frac{\mu_0 i_1 i_2}{2\pi x}(\sqrt{(a^2+x^2)}-x).$$

(*b*) Apply the expression for the force

$$F_x = \frac{\mu_0 i_1 i_2}{4\pi}\oint_1\oint_2 \frac{x_2-x_1}{r^3}\cos\epsilon\, dl_1\, dl_2.$$

(*c*) Find the mutual inductance between two square circuits distance x apart (see problem **6.4**) and apply

$$F_x = i_1 i_2 \frac{\partial M}{\partial x}.$$

7. An infinitely long straight wire carrying a current lies along the central axis of an ideal uniformly wound toroidal coil. Calculate the total magnetic flux through the coil.

Show that the mutual inductance between the toroid and any circuit which passes through it is approximately constant.

8. A copper wire is wound in the form of a long solenoid of radius 1 cm and with 10 turns per cm. It is suspended from one end and the other end is attached to a body of 100 gm weight lying on a table. Find the approximate value of the current which must be passed through the wire in order to lift the body. (Neglect the mass of the wire.)

9. An electron spins about its own axis with an angular momentum of $(\sqrt{3}/2)(h/2\pi)$, where h is Planck's constant. Find the magnetic moment associated with the spin on the assumptions that

(*a*) the charge and mass are uniformly distributed over the surface of a sphere and

(*b*) the charge and mass are uniformly distributed over the volume of a sphere.

10. In a certain mass spectrometer a uniform magnetic field within a spherical volume is obtained by passing a current through a coil wound on the surface of the sphere in such a way that the winding density is proportional to $\sin \theta$, where θ is the polar angle measured from a fixed diameter, the winding being in planes normal to this diameter. Prove that the field is uniform.

7 MOTION OF CHARGED PARTICLES IN ELECTRIC AND MAGNETIC FIELDS

The force on a particle, charge q, moving with speed $\boldsymbol{v}$ under the influence of $\boldsymbol{E}$ and $\boldsymbol{B}$ is given by

$$\boldsymbol{F} = q(\boldsymbol{E}+\boldsymbol{v}+\boldsymbol{B}).$$

The rate at which the particle acquires energy is $\boldsymbol{F} \cdot \boldsymbol{v} = q(\boldsymbol{E} \cdot \boldsymbol{v})$ since $\boldsymbol{v} \times \boldsymbol{B}$ is perpendicular to $\boldsymbol{v}$, Thus the force due to $\boldsymbol{B}$ does not influence the energy of the particle. Note, however, that if $\boldsymbol{B}$ varies with time an electric field is created, given by $\nabla \times \boldsymbol{E} = -\partial \boldsymbol{B}/\partial t$.

7.1. *An electron moves in a region where a uniform electric field* **E** *and a uniform magnetic field of induction* **B** *are at right angles to each other. If the electron is projected with speed v in a direction opposite to* **E** *discuss the motion.*

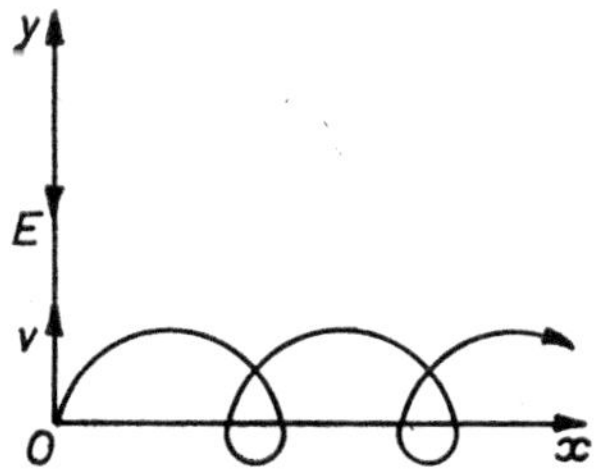

FIG. 36

Let $\boldsymbol{E}$, $\boldsymbol{B}$ be in the negative directions of the y, z axes and let the electron be projected from the origin. Thus the initial conditions are:

$$t = 0 \quad \begin{cases} x = 0 & y = 0, \\ \dot{x} = 0 & \dot{y} = v. \end{cases}$$

The equations of motion are

$$m\ddot{x} = eB\dot{y}, \tag{1}$$

$$m\ddot{y} = eE - eB\dot{x}. \tag{2}$$

Integrating (*1*),

$$m\dot{x} = eBy.$$

Substituting in (*2*),

$$\ddot{y} = \frac{eE}{m} - \frac{e^2B^2}{m^2}y.$$

Let $\omega = eB/m, f = eE/m$, for convenience. Then $\ddot{y}+\omega^2 y = f$.

The solution may be written

$$y = a \sin(\omega t+\phi)+f/\omega^2.$$

From the initial conditions

$$a \sin\phi + \frac{f}{\omega^2} = 0, \quad \omega a \cos\phi = v.$$

Therefore

$$a^2 = \frac{f^2}{\omega^4}+\frac{v^2}{\omega^2} \quad \text{and} \quad \tan\phi = -\frac{f}{v\omega}.$$

or

$$a = \frac{1}{\omega}\left(\frac{E^2}{B^2}+v^2\right)^{\frac{1}{2}}, \quad \tan\phi = -\frac{E}{vB}.$$

We have

$$\dot{x} = \omega y = \omega a \sin(\omega t+\phi)+\frac{f}{\omega},$$

i.e.

$$\dot{x} = \omega a \sin(\omega t+\phi)+\frac{E}{B},$$

$$\dot{y} = \omega a \cos(\omega t+\phi),$$

and

$$x = (E/B)t - a\cos(\omega t+\phi)+a\cos\phi.$$

$$y = a\sin(\omega t+\phi)+(E/\omega B).$$

The motion may thus be described as a rotation in a circle of radius a with constant angular velocity $\omega = eB/m$ about its centre which moves along the line $y = E/\omega B$ with constant speed E/B. The path is a trochoid.

Show that if the electron is initially at rest the path will be a cycloid.

7.2. *A diode valve consists of a straight filament of radius a along the axis of a cylindrical anode of radius b. It is immersed in a magnetic field of induction B parallel to the filament. If the anode has a potential V find the condition that electrons emitted from the filament with negligible velocity should fail to reach the anode.*

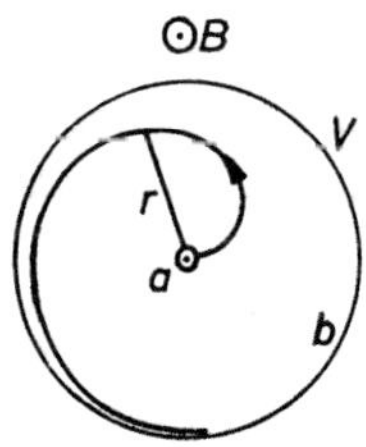

FIG. 37

The motion of an electron will be given (in plane polar coordinates) by

$$m(\ddot{r}-r\dot{\theta}^2) = eE - eBr\dot{\theta}, \qquad (1)$$

$$\frac{m}{r}\frac{d}{dt}(r^2\dot{\theta}) = eB\dot{r}. \qquad (2)$$

Equation (2) gives

$$\frac{d}{dt}(r^2\dot{\theta}) = \frac{eB}{m}r\dot{r} = \frac{eB}{m}\frac{d}{dt}(\tfrac{1}{2}r^2).$$

Integrating,

$$r^2\theta = \frac{eB}{2m}(r^2-a^2).$$

If the electron just grazes the anode it will be moving at right angles to the radius vector ($\dot{r} = 0$) so that its velocity will be $r\dot{\theta}$ and its energy $\frac{1}{2}mr^2\dot{\theta}^2$, for $r = b$. This energy will be equal to eV since only the electric field does work on an electron. Thus

$$eV = \tfrac{1}{2}m\left[\frac{eB}{2m}\frac{(b^2-a^2)}{b}\right]^2.$$

Hence the electron will not reach the anode if

$$V < \frac{e}{m}\frac{B^2}{8}\frac{(b^2-a^2)^2}{b^2}.$$

If $a \ll b$,

$$V < \frac{eB^2b^2}{8m}.$$

The value of E at a point distance r from the axis is $V/(r \log b/a)$ towards the axis. The path of the electron would be given by the integration of equation (*1*), which is not easy.

7.3. *Two large circular conducting plates are mounted parallel to each other* 1 mm *apart in a vacuum. The plates are raised to a potential difference of* 1000 volts *and are immersed in a uniform magnetic field of induction* 0·01 weber m^{-2} *parallel to the surface of the plates. A radioactive source of beta-rays of maximum energy* 1·5 MeV *is situated at the centre of the gap. Find the angular spread in the central plane over which electrons escape from the gap and the energy distribution over this angle, taking the rest-energy of the electron as* 0·5 MeV.

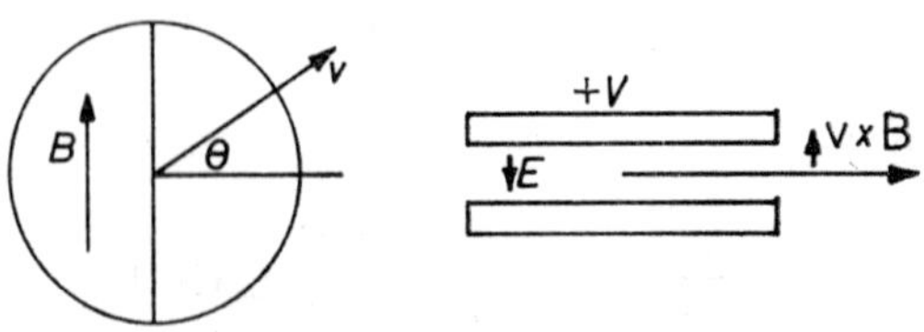

FIG. 38

E, the electric field in the gap $= 10^{-6}$ volt m^{-1}. The condition of escape is that the resultant force on the electron in the gap is zero, i.e. $\boldsymbol{E}+\boldsymbol{v}\times\boldsymbol{B} = 0$. If we consider an electron moving with speed v in the direction shown in the figure, then $eE = eBv\cos\theta$.

Hence

$$v\cos\theta = \frac{E}{B} = \frac{10^6}{10^{-2}} = 10^8 \text{ ms}^{-1}$$

$$= c/3 \quad (c = \text{velocity of light})$$

or

$$\cos\theta = c/3v.$$

For electrons of kinetic energy 1.5 MeV the relativistic energy is 2 MeV. Thus from

$$mc^2 = m_0c^2\left(1-\frac{v^2}{c^2}\right)^{-\frac{1}{2}}$$

we obtain

$$\left(1-\frac{v^2}{c^2}\right)^{-\frac{1}{2}} = 4.$$

Hence the maximum value of $v/c = \sqrt{(15)}/4$. The largest angle will be given by

$$\theta_m = \cos^{-1}\left(\frac{4}{3\sqrt{15}}\right).$$

Electrons will escape for angles between $\pm\theta_m$. The total energy of the electron

$$W = W_0\left(1-\frac{v^2}{c^2}\right)^{-\frac{1}{2}} = \frac{3\cos\theta}{2(9\cos^2\theta-1)^{\frac{1}{2}}}\text{ MeV}.$$

This gives the distribution of W with θ.

For $\theta = 0$, $W = 3\sqrt{(2)}/8$ MeV or kinetic energy $\cong$ 30 keV.

7.4. (*a*) *An electron moves in a circular orbit in a magnetic field of induction* **B** *which is spatially uniform but increases slowly in a time large compared with the period of the electron in its orbit. Show that the momentum of the electron varies as* $\sqrt{B}$ *and the orbit radius as* $1/\sqrt{B}$, *approximately.*

(*b*) *Show that if the rate of change of flux through the orbit is increased to $2\pi r^2\dot{B}$ (as in a betatron) while B remains the magnetic induction at the orbit, then the electron will move in a circle of constant radius and the momentum will be proportional to B.*

(*a*) The conditions of the problem are such that we can assume that the orbit is sensibly circular at all times. Thus at any instant $Bev = mv^2/r$ or the momentum

$$mv = eBr \tag{1}$$

The rate of change of flux through a circle of radius r will be given by $\pi r^2\dot{B}$ which can be equated to $2\pi rE$, where E is the electric field tangential to the circle. Thus

$$E = r\dot{B}/2.$$

The tangential force

$$\frac{d}{dt}(mv) = Ee = er\dot{B}/2. \tag{2}$$

Differentiating (*1*),

$$\frac{d}{dt}(mv) = e(B\dot{r}+r\dot{B}). \tag{3}$$

Hence from (*2*) and (*3*)

$$B\dot{r} = -r\dot{B}/2.$$

$$\frac{1}{B}\frac{dB}{dt} = -\frac{2}{r}\frac{dr}{dt} \quad \text{or} \quad \frac{dB}{B} = -2\frac{dr}{r}.$$

Therefore $B = k/r^2$, where k can be found from the initial conditions, if desired. Hence the radius varies as $1/\sqrt{B}$, and from (*1*) the momentum varies as $\sqrt{B}$.

(*b*) If the rate of change of flux through the orbit is $2\pi r^2\dot{B}$ then $E = r\dot{B}$ and

$$\frac{d}{dt}(mv) = er\dot{B}.$$

By comparison with (*3*) we obtain $\dot{r} = 0$ or $r =$ constant, and from (*1*) mv is proportional to B.

7.5. *A long solenoid is wound with* 20 *turns of wire per cm. Inside the solenoid is placed a vacuum tube with an electron gun which produces a beam through an anode slit at* 10 kV. *The beam passes along the axis of the solenoid and strikes a fluorescent screen* 42 *cm from the anode. When the current flowing in the coil is* 2 A (*or a multiple thereof*) *an image of the anode slit is produced on the screen. Find the value of e/m for electrons.*

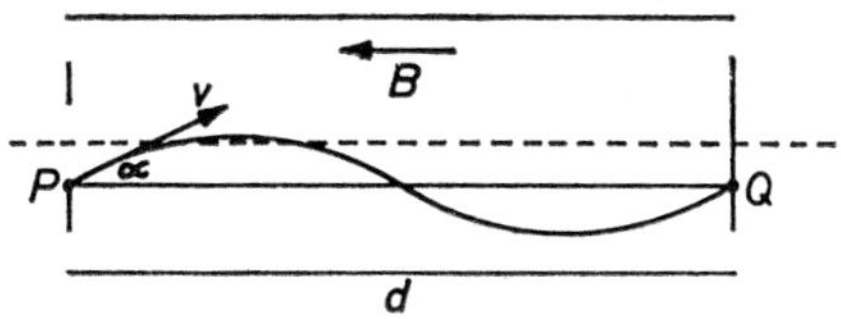

FIG. 39

Consider an electron which leaves the anode slit from the point P with speed v at angle α to the axis of the solenoid. After leaving the anode the electron moves only in the magnetic field under the force $-e(\mathbf{v}\times\mathbf{B})$, of magnitude $evB\sin\alpha$ in a plane normal to the axis. The path will be a helix which projects to a circle on that plane. If r is the radius of this circle $Ber = mv\sin\alpha$.

The time taken to complete the circle is $2\pi r/v\sin\alpha = 2\pi m/Be$, which is independent of α. During this time the electron has travelled to the point Q, where PQ is parallel to the axis of the solenoid, with the constant speed $v\cos\alpha$.

Hence

$$d = \frac{2\pi m}{Be} v\cos\alpha.$$

If the angle α is always small then all the electrons diverging from P will come to a focus on the screen at Q. The same is true for all points of the anode aperture. Thus an image of the slit will appear on the screen, of the same dimensions as the slit.

Hence for α small

$$d = \frac{2\pi mv}{Be}.$$

If V is the anode potential, we have $eV = \frac{1}{2}mv^2$, so that

$$\frac{e}{m} = \frac{8\pi^2 V}{B^2 d^2}.$$

$$B = \mu_0 nI = 4\pi 10^{-7} \times 2000 \times 2 = 16\pi \times 10^{-4} \text{ Wb m}^{-2},$$

which gives

$$\frac{e}{m} = 1{\cdot}77 \times 10^{11} \text{ C kg}^{-1}.$$

7.6. *Show that the magnetic field produced by a current flowing in a short circular coil co-axial with an evacuated tube containing an electron beam behaves like a lens and brings a divergent beam to a focus.*

Find an approximate expression for the focal length of the magnetic lens if the coil has N turns, current I, and is of mean diameter d.

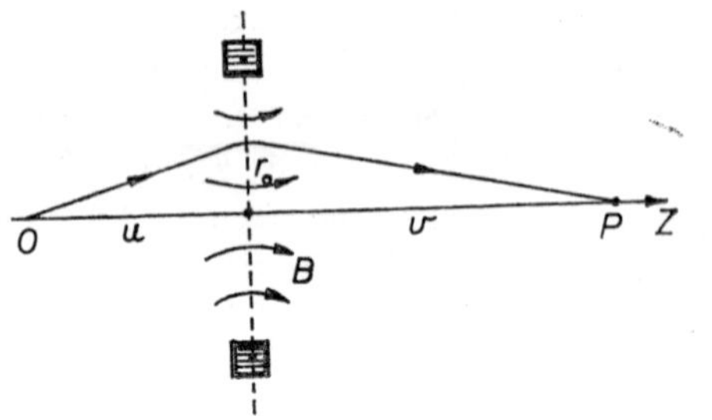

FIG. 40

It is assumed that the region in which the magnetic field exists is short compared with the total length of the electron path.

Let r, ϕ, z be the cylindrical coordinates of the electron position, the z-axis being the axis of the coil. Then at all points $\nabla . \mathbf{B} = 0$, or

$$\frac{1}{r}\frac{\partial}{\partial r}(rB_r) + \frac{1}{r}\frac{\partial}{\partial \phi}B_\phi + \frac{\partial B_z}{\partial z} = 0.$$

Here $\mathbf{B}$ is a function of r, z but not of ϕ.

Hence

$$\frac{1}{r}\frac{\partial}{\partial r}(rB_r) = -\frac{\partial B_z}{\partial z}. \tag{1}$$

We assume that as the electron moves through the field the value of r is small and does not change appreciably. (This assumption is the same as that for a thin lens in optics.) If r has the constant value r_0 within the field we can integrate the above equation and obtain

$$r_0 B_r = -\frac{r_0^2}{2}\frac{\partial B_z}{\partial z} \quad \text{or} \quad B_r = -\frac{r_0}{2}\frac{\partial B_z}{\partial z}. \tag{2}$$

This equation gives the radial component of B in terms of the variation of B_z along the z-axis.

The non-relativistic equations of motion of an electron moving within the field are

$$m(\ddot{r}-r\dot{\phi}^2) = -e(r\dot{\phi}B_z). \tag{3}$$

$$m(r\ddot{\phi}+2\dot{r}\dot{\phi}) = -e(\dot{z}B_r-\dot{r}B_z). \tag{4}$$

$$m\ddot{z} = -e(-r\dot{\phi}B_r). \tag{5}$$

For $r = r_0$ equation (4) becomes

$$r_0\ddot{\phi} = -\frac{e}{m}\dot{z}B_r,$$

or

$$\ddot{\phi} = \frac{e}{2m}\dot{z}\frac{\partial B_z}{\partial z}, \quad \text{using (2)}$$

Thus

$$\dot{\phi} = \frac{eB_z}{2m}. \tag{6}$$

From this relation we see that the electron acquires an angular velocity about the z-axis.

Substituting the above value of $\dot{\phi}$ in equation (3)

$$\ddot{r} = -\frac{e^2 r_0}{4m^2}B_z^2. \tag{7}$$

The velocity of the electrons in the z direction is approximately constant. Substitution for B_r in equation (5) shows that $\ddot{z}$ involves

r_0^2, the square of a small quantity. Treating $\dot{z}$ as constant we can write

$$\frac{d}{dt} = \dot{z}\frac{d}{dz} \quad \text{and also} \quad \tfrac{1}{2}m\dot{z}^2 = Ve,$$

where V is the potential through which the electrons were accelerated at the source.

Thus

$$\frac{d}{dt} = \sqrt{\left(\frac{2eV}{m}\right)}\frac{d}{dz},$$

and equation (7) becomes

$$\frac{2eV}{m}\frac{d^2r}{dz^2} = -\frac{e^2r_0}{4m^2}B_z^2, \quad \text{or} \quad \frac{d^2r}{dz^2} = -\frac{er_0}{8mV}B_z^2. \tag{8}$$

We can integrate this equation through the region of the field if B_z is known as a function of z. Since we have assumed that the range of z for which $B > 0$ is short compared with the total path, the trajectory will be linear before and after the coil. Let $(dr/dz)_1$ be the initial gradient of the path and $(dr/dz)_2$ the final gradient, then

$$\left(\frac{dr}{dz}\right)_2 - \left(\frac{dr}{dz}\right)_1 = -\frac{er_0}{8mV}\int_1^2 B_z^2\,dz.$$

If u is the distance between the source 0 and the plane of the coil (the object distance) and v the distance between the coil and the point P where the electron trajectory intersects the axis (the image distance), then

$$\frac{1}{v}+\frac{1}{u} = \frac{e}{8mV}\int_1^2 B_z^2\,dz.$$

Thus all paraxial electrons emitted from 0 will be focused at the point P. The magnetic field acts like a converging lens of focal length f where

$$\frac{1}{f} = \frac{e}{8mV}\int_1^2 B_z^2\,dz. \tag{9}$$

Such a magnetic lens differs from an optical lens in that the image of an extended source is rotated about the axis through an angle ϕ which is obtained by integrating equation (*6*).

$$\phi = \sqrt{\frac{e}{8mV}} \int_1^2 B_z \, dz. \tag{10}$$

We can obtain approximate values for f and ϕ for a narrow coil if we assume that the field distribution differs little from that of a single turn of current NI and radius a.

$$B_z = \frac{\mu_0 NI}{2} \frac{a^2}{(a^2+z^2)^{\frac{3}{2}}},$$

if z is measured from the plane of the coil.

Although B_z is not confined to a short range of z, its value falls off rapidly with z and no serious error will be introduced by taking the integral between infinite limits.

$$\int_{-\infty}^{\infty} B_z^2 \, dz = \frac{\mu_0^2 N^2 I^2}{4} \int_{-\infty}^{\infty} \frac{a^4 \, dz}{(a^2+z^2)^3}.$$

$$= \frac{\mu_0^2 N^2 I^2}{4} \cdot \frac{3\pi}{8a} \quad \text{(Substitute } z = a \tan \phi.\text{)}$$

Hence

$$\frac{1}{f} = \frac{e}{8mV} \frac{3\pi \mu_0^2 N^2 I^2}{32a}.$$

Substituting

$$e = 1{\cdot}6 \times 10^{-19} \text{ C}, \quad m = 9 \times 10^{-31} \text{ kg}, \quad d = 2a,$$

and expressing V in kilovolts, we find

$$NI \cong 220\sqrt{Vd/f}.$$

The angle of rotation of the image will be given by (*10*).

$$\textstyle\int_{-\infty}^{\infty} B_z \, dz = \mu_0 NI.$$

This can be obtained easily since it is the same integral as that for any closed path threading the coil.

Hence

$$\phi = \sqrt{\left(\frac{e}{8mV}\right)} \mu_0 NI$$

$$\cong (6 \times 10^{-3}) \frac{NI}{\sqrt{V}} \text{ radian.}$$

7.7. *A thin wire is held by a spring under a tension T between two supports in a horizontal plane. It is placed in a uniform vertical magnetic field of induction B. Show that when a current I flows in the wire it assumes the form of an arc of a circle of radius T/BI.*

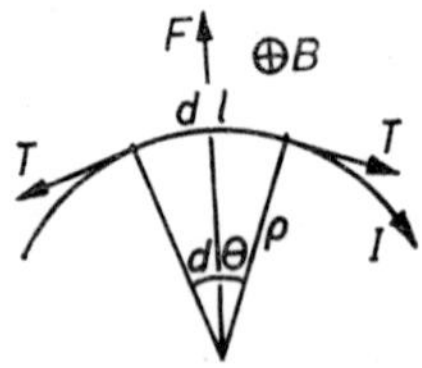

FIG. 41

The wire will assume a position in which the force exerted by the magnetic field on the current is balanced by the tension in the wire. Consider an element dl of the wire. The force due to B is $IB\,dl = F$. If ρ is the radius of curvature and $d\theta$ is the angle subtended at the centre of curvature by dl then $\rho\,d\theta = dl$ and since the element is in equilibrium we have

$$2T \sin\frac{d\theta}{2} = IB\,dl = IB\rho\,d\theta.$$

Hence in the limit $\rho = T/BI$, which is a constant. The wire, if supported freely at its ends within the region of the uniform field, will lie along the arc of a circle of the above radius.

7.8. *A thin wire is supported horizontally under a tension of* 50 gm weight *between two smooth supports P and Q in the central plane of the gap of an electromagnet which provides a uniform induction of* 0.5 Wb m^{-2}. *A current of one ampère is passed through the wire in the direction from P to Q. Find the radius of the circle on which the wire lies and show that this is identical with the trajectory of a charged particle of a defined momentum.*

From the previous problem the radius of the arc of the circle is given by

$$\frac{T}{BI} = \frac{50 \times 10^{-3} \times 9{\cdot}81}{0{\cdot}5} = 0{\cdot}981 \text{ m}.$$

Consider the motion of a charged particle, mass m, charge e in the field. Let the speed of projection be v in the horizontal plane.

The force on the particle $= e(\boldsymbol{v} \times \boldsymbol{B})$. Hence $mv^2/\rho = Bev$, or $mv = Be\rho$. The particle will move in a circle of radius $\rho = mv/Be$. For the wire $B\rho = T/I$ and for the particle $B\rho = mv/e$. Hence for the same circle the momentum of the particle $p = mv = Te/I$ (note that the value of B does not appear).

$$p = \frac{0{\cdot}981}{2} 1{\cdot}6 \times 10^{-19} \text{ kg m s}^{-1}$$

if the particle is singly charged. In nuclear physics it is usual to express momentum in units of MeV/c. 1 electron-volt $= 1{\cdot}6 \times 10^{-19}$ joule.

$$1 \text{ MeV/c} = \frac{1{\cdot}6 \times 10^{-13}}{3 \times 10^8} = \frac{1{\cdot}6}{3} 10^{-21} \text{ kg m s}^{-1},$$

Hence in this case

$$p = \frac{0{\cdot}981 \times 300}{2} = 147 \text{ MeV/c}.$$

Note that in the motion of the particle the force due to B is directed radially inwards while in the case of the wire it is radially outwards. Thus the particle moving from P to Q along the same curve as the wire will be negatively charged.

This problem illustrates a well-known experimental method for mapping the trajectories of charged particles in a magnetic field.

7.9. *A long straight co-axial cable with a vacuum dielectric has a central conductor of diameter* 4 cm *and an outer conductor of internal diameter* 20 cm. *A current of* 4000 A *flows along the internal conductor which has a negligible resistance: a potential difference of* 300 k.V. *is maintained between the conductors. A stream of singly charged particles enters the space between the conductors with a momentum of* 601 MeV/c *directed parallel to the axis. What will be the rest-mass of those particles which continue to move in a direction parallel to the axis?*

If the particles move parallel to the axis then $Ee = Bev$ or $v = E/B$.

If V is the potential difference between the conductors the value of E distance r from the axis is

$$\frac{1}{r}\frac{V}{\log b/a}.$$

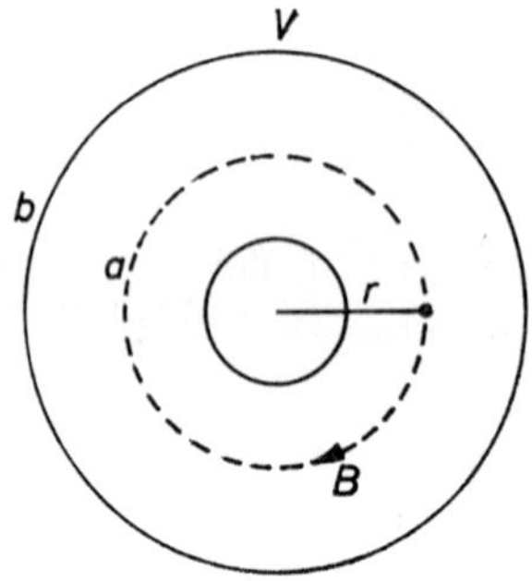

FIG. 42

Let ρ be the charge per unit length on the central conductor. Then

$$E = \frac{\rho}{2\pi\Sigma_0}\frac{1}{r}$$

and

$$V = \int_a^b E\,dr = \frac{\rho}{2\pi\Sigma_0}\int_a^b \frac{dr}{r} = \frac{\rho}{2\pi\Sigma_0}\log\frac{b}{a}.$$

Hence

$$E = \frac{1}{r}\frac{V}{\log b/a}.$$

Since

$$B = \frac{\mu_0 I}{2\pi r}, \quad v = \frac{2\pi V}{\mu_0 I \log b/a}.$$

Thus

$$\frac{v}{c} = \frac{2\pi(3\times 10^5)}{(4\pi 10^{-7})(4000)\log_e 5}\cdot\frac{1}{(3\times 10^8)}$$

$$= 0{\cdot}777. \quad (c = 3\times 10^8\ \mathrm{ms}^{-1}.)$$

The particle is relativistic, hence if W_0 is the rest-energy and W is the total energy, $W = \gamma W_0$ where $\gamma = (1-[v^2/c^2])^{-\frac{1}{2}}$.

The relation between momentum p, W and W_0 is $p^2c^2 = W^2 - W_0^2$ if energy is expressed in joules.

If the energy is measured in MeV (1 MeV $= 1{\cdot}6\times 10^{-13}$ joule) and momentum in MeV/c, then we have

$$\begin{aligned} 601 &= (W^2-W_0^2)^{\frac{1}{2}} = W_0(\gamma^2-1)^{\frac{1}{2}} \\ &= W_0(1{\cdot}234). \\ W_0 &= 487 \text{ MeV}. \end{aligned}$$

This energy is that of a K-meson.

EXERCISES

1. Electrons are liberated with zero velocity from the negative plate of a parallel plate condenser to which is applied a magnetic field of induction B parallel to the plates. Show that the electrons will not reach the positive plate (potential V) if the plate separation is greater than

$$\frac{1}{B}\sqrt{\frac{2mV}{e}}.$$

2. A long straight filament of radius a carries a current I and emits electrons which are accelerated towards a long concentric cylindrical anode of radius b, raised to a potential V above the filament. If the initial velocity of the electrons is negligible show that they fail to reach the anode if

$$V < \frac{\mu_0^2 I^2 e}{8\pi^2 m}\left(\log\frac{b}{a}\right)^2.$$

3. A particle of charge e, rest-mass m_0, is projected from the origin along the y-axis in a uniform electric field parallel to the x-axis. If the variation of mass with velocity is taken into account, find x and y as functions of time and show that the path of the particle is

$$x = \frac{W_1}{eE}\left(\cosh\frac{eEy}{p_1 c} - 1\right)$$

where $W_1 = (m_0^2c^4 + p_1^2c^2)^{\frac{1}{2}}$ and p_1 is the initial momentum.

4. A charged particle of mass m is projected from the origin with velocity v_0 in a direction making an angle θ with the z-axis. A uniform magnetic field of induction B and a uniform electric field E, both directed in the negative z-direction, act on the particle. Show that the particle returns to the origin if $v_0 = \pi E/B \cos\theta$. (Neglect relativistic effects.)

5. An electron synchrotron has an orbit radius of 1·25 m and the magnetic field is one tesla. Find the energy of the circulating electrons in MeV and show that their speed differs from that of light by approximately one part in a million.

6. Any axially symmetrical electric field will produce the effect of a lens on an electron beam passing through it. Show that if V is the potential at a point on the axis (taken as the z-axis) and r is the radial displacement of the electron from the axis, the differential equation of the trajectory of paraxial electrons is

$$\sqrt{V}\frac{d}{dz}\left(\sqrt{V}\frac{dr}{dz}\right)+\frac{r}{4}\frac{d^2V}{dz^2} = 0.$$

7. Within a quadrupole magnet the magnetic field distribution is given by

$$B_x = ky, \quad B_y = kx, \quad B_z = 0; \quad 0 < z < l,$$

where k is a constant. A parallel beam of electrons of velocity v in the z-direction enters the field at the origin. Show that the beam will be convergent in the yz-plane and divergent in the xz-plane, the equivalent focal lengths being given numerically by $(\omega \sin \omega l)^{-1}$ and $(\omega \sinh \omega l)^{-1}$, respectively, where $\omega^2 = ke/mv$ and $\omega l < \pi/2$.

8 ELECTROMAGNETIC WAVES

The general equations of the electromagnetic field are

$$\nabla \times \boldsymbol{E} = -\frac{\partial \boldsymbol{B}}{\partial t}, \quad \nabla . \boldsymbol{D} = \rho, \quad \boldsymbol{D} = \epsilon \boldsymbol{E},$$

$$\nabla \times \boldsymbol{H} = \boldsymbol{j} + \frac{\partial \boldsymbol{D}}{\partial t}, \quad \nabla . \boldsymbol{B} = 0, \quad \boldsymbol{B} = \mu \boldsymbol{H}.$$

$\boldsymbol{j} = \sigma \boldsymbol{E}$ in a conductor or $\boldsymbol{j} = \rho \boldsymbol{v}$ for distributed charge. In a uniform medium with $\boldsymbol{j} = 0$, $\rho = 0$,

$$\nabla \times \boldsymbol{E} = -\mu \frac{\partial \boldsymbol{H}}{\partial t}; \quad \nabla \times \boldsymbol{H} = \epsilon \frac{\partial \boldsymbol{E}}{\partial t}.$$

$$\nabla \times (\nabla \times \boldsymbol{E}) = \nabla(\nabla . \boldsymbol{E}) - \nabla^2 \boldsymbol{E}$$

$$= -\frac{\partial}{\partial t}(\nabla \times \boldsymbol{B}) = -\mu\epsilon \frac{\partial^2 \boldsymbol{E}}{\partial t^2}.$$

Since $\nabla . \boldsymbol{E} = 0$ we have the wave equation

$$\nabla^2 \boldsymbol{E} = \mu\epsilon \frac{\partial^2 \boldsymbol{E}}{\partial t^2}.$$

Similarly

$$\nabla^2 \boldsymbol{H} = \mu\epsilon \frac{d^2 \boldsymbol{H}}{\partial t^2}.$$

Velocity of propagation

$$c = 1/\sqrt{\mu\epsilon}$$

In free space

$$c = 1/\sqrt{\mu_0 \epsilon_0} = 3 \times 10^8 \text{ ms}^{-1}.$$

Plane waves. If the direction of propagation is $\boldsymbol{n}$ (d . cos l, m, n) all components of $\boldsymbol{E}$ and $\boldsymbol{H}$ can be taken as

$$\text{(const.)} \quad \exp\left[i\omega\left(t - \frac{lx + my + nz}{c}\right)\right]$$

where ω can have any value.

Substituting in Maxwell's equations we find

$$\sqrt{\mu}(\boldsymbol{n}\times\boldsymbol{H}) = -\sqrt{\epsilon}\boldsymbol{E};\quad \sqrt{\epsilon}(\boldsymbol{n}\times\boldsymbol{E}) = \sqrt{\mu}\boldsymbol{H},$$

or $\boldsymbol{E}$, $\boldsymbol{H}$, $\boldsymbol{n}$ form a right-handed set of axes and $\sqrt{\epsilon}\boldsymbol{E} = \sqrt{\mu}H$.

Poynting's vector. The rate of flow of energy through any surface $= \int \boldsymbol{P}\,.\,d\boldsymbol{S}$ where $\boldsymbol{P} = \boldsymbol{E}\times\boldsymbol{H}$ (Wm^{-2}). For plane waves the mean value of

$$\boldsymbol{P} = \overline{\boldsymbol{P}} = \sqrt{\frac{\epsilon}{\mu}}\,E^2 = E^2/Z,$$

where E is the r.m.s. value of the electric field and $Z = \sqrt{\mu/\epsilon}$ is the wave impedance of the medium. For free space $Z_0 = 377\,\Omega$. Mean energy density

$$\overline{W} = \tfrac{1}{2}(\epsilon E^2+\mu H^2) = \epsilon E^2\ (\mathrm{Jm}^{-3})$$

Hence

$$\overline{P} = \overline{W}\times c.$$

Mean momentum density $= (1/c^2)\,\overline{[\boldsymbol{E}\times\boldsymbol{H}]} = \overline{W}/c$. Pressure exerted by waves completely absorbed in a surface $= \overline{W} = \epsilon E^2$ (Nm^{-2}). Conducting medium: Here $\boldsymbol{j} = \sigma\boldsymbol{E}$

$$\nabla\times\boldsymbol{H} = \sigma\boldsymbol{E}+\Sigma\frac{\partial\boldsymbol{E}}{\partial t};\quad \nabla\times\boldsymbol{E} = -\mu\frac{\partial H}{\partial t}.$$

Equation of propagation

$$\nabla^2\boldsymbol{E} = \mu\epsilon\frac{\partial^2\boldsymbol{E}}{\partial t^2}+\mu\sigma\frac{\partial\boldsymbol{E}}{\partial t}.$$

with a similar equation for $\boldsymbol{H}$.

Assume a solution

$$\exp\left[i\omega\left(t-\frac{lx+my+nz}{v}\right)\right].$$

On substitution

$$\frac{1}{v^2} = \mu\epsilon\left(1-i\frac{\sigma}{\omega\epsilon}\right).$$

Let

$$\frac{1}{\sqrt{(\mu\epsilon)}v} = \left[1 - i\frac{\sigma}{\omega\epsilon}\right]^{\frac{1}{2}} = \alpha + i\beta. \quad (\alpha, \beta \text{ real})$$

Then

$$\alpha = +[\tfrac{1}{2}(\sqrt{[1+\sigma^2/\omega^2\epsilon^2]}+1)]^{\frac{1}{2}}; \quad \beta = -[\tfrac{1}{2}(\sqrt{[1+\sigma^2/\omega^2\epsilon^2]}-1]^{\frac{1}{2}},$$

and the propagation is given by

$$\exp\{i\omega[t-(lx+my+nz)\sqrt{\mu\epsilon\alpha}]\}\exp(\omega\beta\sqrt{(\mu\epsilon)}z).$$

Thus the introduction of σ leads to the results

(*a*) Phase velocity $= 1/\alpha\sqrt{\mu\epsilon}$,

(*b*) Amplitude attenuation e^{-kz} where $k = -\omega\beta\sqrt{\mu\epsilon}$,

(*c*) $\boldsymbol{H}$ and $\boldsymbol{E}$ are not in time phase.

$$\boldsymbol{H} = \frac{1}{v\mu}(\boldsymbol{n}\times\boldsymbol{E}) = \sqrt{\frac{\epsilon}{\mu}}(\alpha+i\beta).$$

$$H = \sqrt{\frac{\epsilon}{\mu}}(\alpha+i\beta)E = \sqrt{\frac{\epsilon}{\mu}}\sqrt{\alpha^2+\beta^2}\,e^{i\delta}E; \quad \tan\delta = \beta/\alpha.$$

For σ large, $\alpha \approx \sqrt{\sigma/2\omega\epsilon}$ and $\beta = -\alpha$. Phase velocity $= \sqrt{2\omega/\mu\sigma}$. Attenuation constant $= \omega\sqrt{\mu\epsilon}\sqrt{\sigma/2\omega\epsilon} = \sqrt{\sigma\mu\omega/z} = \sqrt{\pi\sigma\mu f}$, where $f =$ frequency in Hz. The amplitude diminishes by a factor of $1/e$ in a distance $(\pi\sigma\mu f)^{-\frac{1}{2}}$. Phase difference between $\boldsymbol{H}$ and $\boldsymbol{E}$, $\delta = -\pi/4$.

For σ small $\beta \approx -\sigma/2\omega\epsilon$ and

$$k = \frac{\sigma}{z}\sqrt{\frac{\mu}{\epsilon}}.$$

If the conductivity is due to the presence of free electrons in an ionised gas $\sigma = -i(Ne^2/m\omega)$ where N is the number of electrons per unit volume and e, m are the charge and mass of the electron.

$$\frac{1}{\sqrt{(\mu\epsilon)}v} = \left(1 - \frac{Ne^2}{m\omega^2\epsilon}\right)^{\frac{1}{2}}.$$

v is real and gives the phase velocity. For ionised air

$$\frac{v}{c} = \left(1-\frac{Ne^2}{m\omega^2\epsilon_0}\right)^{-\frac{1}{2}} \quad (v > c)$$

or the index of refraction

$$= \left(1-\frac{Ne^2}{m\omega^2\epsilon_0}\right)^{+\frac{1}{2}},$$

$$\cong \left(1-\frac{81N}{f^2}\right)^{\frac{1}{2}}.$$

Reflection and Refraction of Electromagnetic Waves

Since they occur in many problems we shall briefly derive here the basic equations which give the electric and magnetic fields in the reflected and refracted waves in terms of the fields in the incident wave on a plane surface of separation of two media.

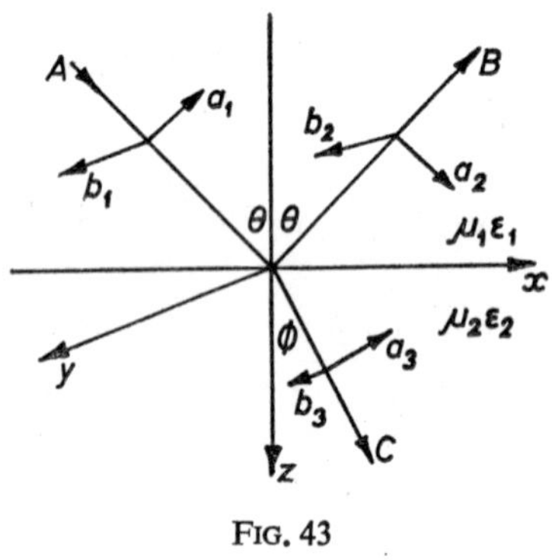

FIG. 43

Let the permittivities and permeabilities of the media be ϵ_1, ϵ_2, μ_1, μ_2 and wave impedances $Z_1 = \sqrt{\mu_1/\epsilon_1}$, $Z_2 = \sqrt{\mu_2/\epsilon_2}$. Let the incident wave (direction AO) have electric field components a_1 in the plane of incidence, b_1 normal to that plane. For the reflected (OB) and refracted (OC) waves we use a_2, b_2, a_3, b_3 in the directions shown in the figure.

The components of the magnetic field can be found from the general relation $\sqrt{\mu}\boldsymbol{H} = \sqrt{\epsilon}(\hat{\boldsymbol{n}}\times\boldsymbol{E})$ where $\hat{\boldsymbol{n}}$ is the direction of propagation, or from $\nabla\times\boldsymbol{E} = -\mu\,\partial\boldsymbol{H}/\partial t$.

The components may be tabulated:

Wave	E	H	*Propagation function*
Incident	$a_1 \cos\theta$ b_1 $-a_1 \sin\theta$	$-b_1 \cos\theta/Z_1$ a_1/Z_1 $b_1 \sin\theta/Z_1$	$\exp\left[i\omega\left(t-\frac{x\sin\theta+Z\cos\theta}{c_1}\right)\right]$
Reflected	$a_2 \cos\theta$ b_2 $a_2 \sin\theta$	$b_2 \cos\theta/Z_1$ $-a_2/Z_1$ $b_2 \sin\theta/Z_1$	$\exp\left[i\omega\left(t-\frac{x\sin\theta-Z\cos\theta}{c_1}\right)\right]$
Refracted	$a_3 \cos\phi$ b_3 $-a_3 \sin\phi$	$-b_3 \cos\phi/Z_2$ a_3/Z_2 $b_3 \sin\phi/Z_2$	$\exp\left[i\omega\left(t-\frac{x\sin\phi+Z\cos\phi}{c_2}\right)\right]$

Index of refraction of second medium relative to first

$$= \frac{\sin\theta}{\sin\phi} = \frac{c_1}{c_2},$$

where

$$c_1 = 1/\sqrt{\mu_1\epsilon_1} \quad c_2 = 1/\sqrt{\mu_2\epsilon_2}.$$

Apply the boundary conditions. It will be sufficient to equate the tangential components of E and H on the two sides of the surface $z = 0$.

$$(E_x) \qquad (a_1+a_2)\cos\theta = a_3\cos\phi$$

$$(H_y) \qquad (a_1-a_2)/Z_1 = a_3/Z_2$$

$$(E_y) \qquad b_1+b_2 = b_3$$

$$(H_x) \quad (b_1-b_2)\cos\theta/Z_1 = b_3\cos\phi/Z_2.$$

These equations give

$$a_2 = -a_1\frac{Z_1\cos\theta-Z_2\cos\phi}{Z_1\cos\theta+Z_2\cos\phi}, \quad a_3 = a_1\frac{2Z_2\cos\theta}{Z_1\cos\theta+Z_2\cos\phi},$$

$$b_2 = -b_1\frac{Z_1\cos\phi-Z_2\cos\theta}{Z_1\cos\phi+Z_2\cos\theta}, \quad b_3 = b_1\frac{2Z_2\cos\theta}{Z_1\cos\phi+Z_2\cos\theta}.$$

If the media are dielectric (non-magnetic) $\mu_1 = \mu_2 = \mu_0$ and

$$\frac{\sin\theta}{\sin\phi} = \sqrt{\frac{\epsilon_2}{\epsilon_1}} = \frac{Z_1}{Z_2}.$$

The above relations become

$$a_2 = -a_1 \frac{\tan(\theta-\phi)}{\tan(\theta+\phi)}, \quad a_3 = a_1 \frac{2\cos\theta\sin\phi}{\sin(\theta+\phi)\cos(\theta-\phi)},$$

$$b_2 = -b_1 \frac{\sin(\theta-\phi)}{\sin(\theta+\phi)}, \quad b_3 = b_1 \frac{2\cos\theta\sin\phi}{\sin(\theta+\phi)}.$$

It is useful to consider normal incidence separately. Here $\theta = 0$, $\phi = 0$ and we need consider only one component.

At the boundary $a_1+a_2 = a_3$, $a_1-a_2 = a_3Z_1/Z_2 = a_3\nu$. Hence

$$a_2 = -a_1 \frac{\nu-1}{\nu+1}, \quad a_3 = a_1 \frac{2}{\nu+1},$$

if $\nu > 1$ the reflected wave is reversed in phase.

R = reflection coefficient = fraction of intensity reflected

$$= \left(\frac{\nu-1}{\nu+1}\right)^2$$

Fraction transmitted

$$= 1-R = \frac{4\nu}{(\nu+1)^2}.$$

Note that the intensity in the second medium is

$$\sqrt{\frac{\epsilon_2}{\mu_0}}\, a_3^2$$

and in the first medium

$$\sqrt{\frac{\epsilon_1}{\mu_0}}\, a_1^2.$$

The fraction transmitted is thus

$$\sqrt{\frac{\epsilon_2}{\epsilon_1}}\,\frac{a_3^2}{a_1^2} = \nu a_3^2/a_1^2.$$

If the incident wave is travelling in the denser medium and ν is the refractive index of the first medium relative to the second we have

$$a_1+a_2 = a_3, \quad \nu(a_1-a_2) = a_3,$$

giving

$$a_2 = a_1 \frac{\nu-1}{\nu+1}, \quad a_3 = \frac{2\nu}{\nu+1} a_1.$$

There is no phase reversal; R has the same value as before, and $a_3 = a_1(1+\sqrt{R})$.

8.1. *A plane electromagnetic wave in air is incident at angle θ on the plane surface of a medium which has relative permeability $\bar{\mu}$ and relative permittivity K. Find the value of θ for which the reflected wave is plane polarised.*

Here $Z_1 = \sqrt{\mu_0/\epsilon_0}$ and $Z_2 = \sqrt{\mu_0\bar{\mu}/\epsilon_0 K}$. Therefore $Z_2/Z_1 = \sqrt{\bar{\mu}/K}$. If ϕ is the angle of refraction we have the relation between a_1 and a_2,

$$a_2 = -a_1 \frac{Z_1 \cos\theta - Z_2 \cos\phi}{Z_1 \cos\theta + Z_2 \cos\phi}.$$

Thus $a_2 = 0$ if $\cos\theta = (Z_2/Z_1)\cos\phi$. But the index of refraction

$$= \frac{\sin\theta}{\sin\phi} = \sqrt{\frac{\mu\epsilon}{\mu_0\epsilon_0}} = \sqrt{\bar{\mu}K}.$$

$$\cos^2\theta = \frac{\bar{\mu}}{K}(1-\sin^2\phi) = \frac{\bar{\mu}}{K}\left(1-\frac{\sin^2\theta}{K\bar{\mu}}\right),$$

which yields

$$\cos^2\theta = \frac{K\bar{\mu}-1}{K^2-1},$$

or

$$\tan^2\theta = \frac{K(K-\bar{\mu})}{K\bar{\mu}-1},$$

provided $K > \bar{\mu}$.

Hence for the value of θ given by this equation, the reflected wave is plane polarised with the electric field normal to plane of incidence. If $K < \bar{\mu}$, $b_2 = 0$, if $\cos\theta = (Z_1/Z_2)\cos\phi = \sqrt{K/\bar{\mu}}\cos\phi$ or

$$\cos^2\theta = \frac{K}{\bar{\mu}}\left(1-\frac{\sin^2\theta}{K\bar{\mu}}\right).$$

Hence

$$\cos^2\theta = \frac{K\bar{\mu}-1}{\bar{\mu}^2-1} \quad \text{or} \quad \tan^2\theta = \frac{\bar{\mu}(\bar{\mu}-K)}{K\bar{\mu}-1}.$$

The reflected wave is polarised with the electric field parallel to the plane of incidence.

8.2. *A plane polarised monochromatic electromagnetic wave is incident at an angle of 45° on the plane surface of separation of two dielectric media and is totally reflected. Show that the reflected wave is elliptically polarised with a phase difference* $\tan^{-1}[n(n^2-2)^{\frac{1}{2}}]$ *where n is the index of refraction of the first medium relative to the second.*

In dealing with problems relating to total reflection we take the usual expressions for a_2 and b_2 but treat $\cos\phi$ as an imaginary quantity.

We have

$$a_2 = -a_1\frac{Z_1\cos\theta - Z_2\cos\phi}{Z_1\cos\theta + Z_2\cos\phi}, \quad b_2 = -b_1\frac{Z_1\cos\phi - Z_2\cos\theta}{Z_1\cos\phi + Z_2\cos\theta}.$$

Here

$$n = \frac{\sin\phi}{\sin\theta} = \frac{c_2}{c_1} = \sqrt{\frac{\epsilon_1}{\epsilon_2}} = \frac{Z_2}{Z_1}.$$

(assuming media are non-magnetic).

Therefore

$$a_2 = -a_1\frac{\cos\theta - n\cos\phi}{\cos\theta + n\cos\phi}, \quad b_2 = -b_1\frac{\cos\phi - n\cos\theta}{\cos\phi + n\cos\theta}.$$

$$\cos\phi = (1-\sin^2\phi)^{\frac{1}{2}} = (1-n^2\sin^2\theta)^{\frac{1}{2}} = i(n^2\sin^2\theta - 1)^{\frac{1}{2}},$$

since for total reflection $n\sin\theta > 1$.

Using $\theta = \frac{\pi}{4}$, $\quad \cos\phi = \frac{i}{\sqrt{2}}(n^2-2)^{\frac{1}{2}}.$

$$a_2 = -a_1\frac{1-in(n^2-2)^{\frac{1}{2}}}{1+in(n^2-2)^{\frac{1}{2}}}, \quad b_2 = b_1\frac{n-i(n^2-2)^{\frac{1}{2}}}{n+i(n^2-2)^{\frac{1}{2}}}.$$

$$= -a_1 e^{-i\delta_1} \qquad\qquad = b_1 e^{-i\delta_2}$$

If

$$e^{-i\delta} = \frac{\alpha - i\beta}{\alpha + i\beta}, \quad \tan \delta/2 = \beta/\alpha.$$

Thus

$$\tan(\delta_1/2) = n(n^2-2)^{\frac{1}{2}}, \quad \tan(\delta_2/2) = (n^2-2)^{\frac{1}{2}}/n.$$

$$\tan(\delta_1 - \delta_2)/2 = \frac{1}{n}(n^2-2)^{\frac{1}{2}}.$$

Hence

$$\tan(\delta_1 - \delta_2) = n(n^2-2)^{\frac{1}{2}}.$$

For the particular angle of incidence chosen, $\delta_2 = \delta_1/2$.

8.3. *A plane polarised radio wave is totally reflected from the surface of water in contact with air. Find the conditions of incidence for which the reflected wave will be circularly polarised. The dielectric constant of water for radio waves is* 81.

For circular polarisation we must have the amplitudes a_2 and b_2 equal in magnitude and 90° different in time phase. Hence, since total reflection does not alter the magnitudes of the field components we must also have $a_1 = b_1$, i.e. the incident wave must be polarised at 45° to the plane of incidence.

From the previous problem

$$a_2 = -a_1 \frac{\cos\theta - in\sqrt{n^2 \sin^2\theta - 1}}{\cos\theta + in\sqrt{n^2\sin^2\theta - 1}},$$

$$= -a_1 e^{-i\delta_1}.$$

$$b_2 = b_1 \frac{n\cos\theta - i\sqrt{n^2\sin^2\theta - 1}}{n\cos\theta + i\sqrt{n^2\sin^2\theta - 1}},$$

$$= b_1 e^{-i\delta_2}.$$

$$\tan\delta_1/2 = \frac{n\sqrt{n^2\sin^2\theta - 1}}{\cos\theta}, \quad \tan\delta_2/2 = \frac{\sqrt{n^2\sin^2\theta - 1}}{n\cos\theta},$$

$$\tan\delta/2 = \tan\frac{\delta_1 - \delta_2}{2} = \frac{\cos\theta\sqrt{n^2\sin^2\theta - 1}}{n\sin^2\theta}.$$

Here

$$\delta = 90°, \quad \tan \delta/2 = 1.$$

$$n^2 \sin^4 \theta = \cos^2 \theta(n^2 \sin^2 \theta - 1)$$

or

$$2n^2 \sin^4 \theta - (n^2+1) \sin^2 \theta + 1 = 0.$$

Since $n = \sqrt{81} = 9$, we have

$$162 \sin^4 \theta - 82 \sin^2 \theta + 1 = 0.$$

This provides two angles of incidence; $\theta = 6°26'$ or $\theta = 44°39'$.

8.4. *A plate of width d of a dielectric medium of index of refraction* ν_2 *separates two dielectric media of indices* ν_1 *and* ν_3*. Monochromatic waves fall normally on the plate. Find the ratio of the amplitude of the reflected wave to that of the incident.*

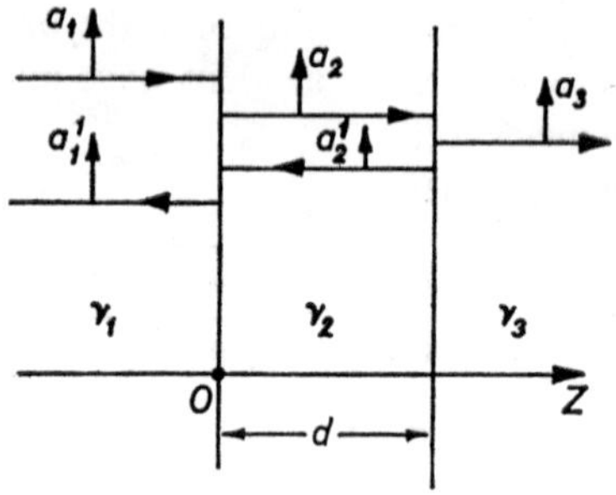

FIG. 44

Take the z-axis normal to the plate. Let a_1 be the electric field in the incident wave and a' that of the reflected wave from the first surface. In the second medium we have a wave in the positive direction of the z-axis and one reflected from the second surface, fields a_2 and a_2', respectively. In the third medium we have a wave in the positive direction, field a_3.

The electric and magnetic fields will be given as below:

Medium	*E*	*H*	
1	a_1	a_1/Z_1	$\exp\left[i\omega\left(t-\frac{z}{c_1}\right)\right]$
	a_1'	$-a_1'/Z_1$	$\exp\left[i\omega\left(t+\frac{z}{c_1}\right)\right]$
2	a_2	a_2/Z_2	$\exp\left[i\omega\left(t-\frac{z}{c_2}\right)\right]$
	a_2'	$-a_2'/Z_2$	$\exp\left[i\omega\left(t+\frac{z}{c_2}\right)\right]$
3	a_3	a_3/Z_3	$\exp\left[i\omega\left(t-\frac{z}{c_3}\right)\right]$

Applying the boundary conditions at the two surfaces, we have, since the media are non-magnetic,

$z = 0$: $$a_1+a_1' = a_2+a_2'. \tag{1}$$

$$(a_1-a_1')v_1 = (a_2-a_2')v_2. \tag{2}$$

$z = d$: $$a_2\exp(-i\omega d/c_2)+a_2'\exp(i\omega d/c_2) = a_3\exp(-i\omega d/c_3). \tag{3}$$

$$[a_2\exp(-i\omega d/c_2)-a_2'\exp(i\omega d/c_2)]v_2 = [a_3\exp(-i\omega d/c_3)]v_3. \tag{4}$$

Assume $v_3 > v_2 > v_1$.

R_1 = reflection coefficient of first surface $= \left(\frac{v_2-v_1}{v_2+v_1}\right)^2$.

R_2 = reflection coefficient of second surface $= \left(\frac{v_3-v_2}{v_3+v_2}\right)^2$.

From (*1*) and (*2*)

$$\frac{a_1'}{a_1} = \frac{a_2(v_1-v_2)+a_2'(v_1+v_2)}{a_2(v_1+v_2)+a_2'(v_1-v_2)} = \frac{-\sqrt{(R_1)}+a_2'/a_2}{1-\sqrt{(R_1)}(a_2'/a_2)}.$$

From (3) and (4), writing $\phi = \omega\, d/c_2$,

$$\frac{a_2'}{a_2} = -\left(\frac{\nu_3-\nu_2}{\nu_3+\nu_2}\right)e^{-i2\phi} = -\sqrt{R_2}\,e^{-i2\phi}.$$

Therefore

$$\frac{a_1'}{a_1} = -\frac{\sqrt{R_1}+\sqrt{R_2}\,e^{-i2\phi}}{1+\sqrt{R_1R_2}\,e^{-i2\phi}}.$$

8.5. *Find the conditions for which there will be no reflection from the plate in the previous problem.*

If $a_1' = 0$, then $\sqrt{R_1}+\sqrt{R_2}\,e^{-i2\phi} = 0$.

$$\text{Let the thickness } d = (2n+1)\frac{\lambda}{4}$$

where n is an integer and λ is the wavelength in the second medium. $\omega = 2\pi c_2/\lambda$.

Then $2\phi = (2n+1)\pi$ which gives $e^{-i2\phi} = -1$.

Hence

$$R_1 = R_2, \quad \text{or} \quad \frac{\nu_2-\nu_1}{\nu_2+\nu_1} = \frac{\nu_3-\nu_2}{\nu_3+\nu_2},$$

i.e.

$$\nu_2^2 = \nu_1\nu_3.$$

8.6. *A beam of monochromatic light in vacuo, angular frequency ω, is incident normally on a film of dielectric of thickness d and refractive index ν. Show that the coefficient of reflection of the film is*

$$\frac{2R(1-\cos 2\phi)}{1+R^2-2R\cos 2\phi}$$

where R is the coefficient of reflection of one surface of the film and $\phi = \omega\nu\, d/c$.

Referring to the solution of **8.4** we see that $\nu_3 = \nu_1 = 1$ and

$$R = \left(\frac{\nu-1}{\nu+1}\right)^2.$$

Thus

$$\frac{a_2'}{a_2} = \left(\frac{\nu-1}{\nu+1}\right) e^{-2i\phi} = \sqrt{R}\, e^{-2i\phi}.$$

This leads to

$$\frac{a_1'}{a_1} = -\sqrt{R}\,\frac{1-e^{-2i\phi}}{1-R\,e^{-2i\phi}}.$$

Therefore

$$\left|\frac{a_1'}{a_1}\right|^2 = R\,\frac{(1-\cos 2\phi)^2+(\sin 2\phi)^2}{(1-R\cos 2\phi)^2+(R\sin 2\phi)^2},$$

$$= 2R\,\frac{1-\cos 2\phi}{1+R^2-2R\cos 2\phi}.$$

Alternative solution. Problems of this type can also be solved by the method of successive reflections. Let $\alpha = (\nu-1)/(\nu+1) = \sqrt{R}$ and, for convenience, let the incident electric field be of unit magnitude.

At the first surface we have $-\alpha$ reflected and $1-\alpha$ transmitted. At the second surface we have $(1-\alpha)\,e^{-i\phi}$ incident, $\alpha(1-\alpha)\,e^{-i\phi}$ reflected and $(1-\alpha^2)\,e^{-i\phi}$ transmitted. (There is no phase change at reflection from a less dense medium and the amplitude of the transmitted wave is $(1+\alpha)$ times the incident.)

Returning to the first surface the next contribution to the first medium is $\alpha(1-\alpha^2)\,e^{-2i\phi}$. Subsequent contributions will be diminished successively by the factor $\alpha^2\,e^{-2i\phi}$ because between them there are two reflections and a path length of $2d$.

Hence the total amplitude reflected from the first surface is

$$-\alpha+\alpha(1-\alpha^2)\,e^{-2i\phi}+\alpha^3(1-\alpha^2)\,e^{-4i\phi}+\ldots$$

$$= -\alpha+\alpha(1-\alpha^2)\,e^{-2i\phi}(1+\alpha^2\,e^{-2i\phi}+\alpha^4\,e^{-4i\phi}\ldots),$$

$$= -\alpha+\frac{\alpha(1-\alpha^2)\,e^{-2i\phi}}{1-\alpha^2\,e^{-2i\phi}},$$

$$= -\frac{\alpha-\alpha\,e^{-2i\phi}}{1-\alpha^2\,e^{-2i\phi}} = -\sqrt{R}\,\frac{1-e^{-2i\phi}}{1-R\,e^{-2i\phi}}.$$

By similar reasoning the amplitude of the field in the transmitted wave is

$$(1-\alpha^2)\,e^{-i\phi}+\alpha^2(1-\alpha^2)\,e^{-3i\phi}+\ldots$$
$$=\frac{(1-\alpha^2)\,e^{-i\phi}}{1-\alpha^2\,e^{-2i\phi}}=\frac{(1-R)\,e^{-i\phi}}{1-R\,e^{-2i\phi}}.$$

The transmitted intensity

$$=\frac{(1-R)^2}{1+R^2-2R\cos 2\phi}.$$

The sum of the reflected and transmitted intensities is, of course, unity.

8.7. *Compare the forces due to gravitational attraction and to radiation pressure from the sun on a black spherical particle of specific gravity 5, and estimate from the following data the radius of the particle for which these forces balance.*

Mean distance of earth from sun $= 1\cdot 5\times 10^{11}$ m.
Period of rotation of earth round sun $= 3\cdot 15\times 10^{7}$ sec.
Intensity of sun's radiation at surface of earth $= 1\cdot 4$ kW m^{-2}.

Let the sun have mass M, total radiated power S.

Let the small sphere have mass m, density ρ, radius r, and be distant d from the sun.

The pressure of radiation on a black body is equal to the energy density which is given by the radiation intensity divided by velocity of propagation.

Radiation pressure on sphere

$$=\frac{1}{c}\cdot\frac{S}{4\pi d^2}\cdot\pi r^2$$

Gravitational attraction

$$=G\frac{Mm}{d^2}\quad (G = \text{constant of gravitation})$$

For balance

$$GMm = Sr^2/4c$$

i.e.

$$GM(\tfrac{4}{3}\pi r^3\rho) = Sr^2/4c$$

$$\therefore \quad r = \frac{3}{16\pi}\frac{S}{GM\rho c}.$$

Let D = earth's distance from sun, v = earth's speed in orbit, T = period of orbit.

Then

$$\frac{GM}{D^2} = \frac{v^2}{D} = \frac{4\pi^2 D}{T^2} \quad \text{or} \quad GM = \frac{4\pi^2 D^3}{T^2}.$$

$$r = \frac{3}{16\pi}\frac{S}{\rho c}\frac{T^2}{4\pi^2 D^3} = \frac{3}{16\pi^2}\frac{I}{\rho c}\frac{T^2}{D},$$

where

$$I = \frac{S}{4\pi D^2} = 1\cdot 4\times 10^3 \text{ W m}^{-2}.$$

Hence

$$r = \frac{3}{16\pi^2}\frac{(1\cdot 4\times 10^3)(3\cdot 5\times 10^7)^2}{(5\times 10^3)(3\times 10^8)(1\cdot 5\times 10^{11})} \text{ m},$$

$$= 1\cdot 16\times 10^{-7} \text{ m}.$$

8.8. *A linear electric quadrupole consists of two electric charges $+Q$ a small distance $2l$ apart with a charge $-2Q$ at the mid-point of the line joining them. If Q varies with time according to $Q = Q_0 \sin \omega t$ find the radiative magnetic field at a distant point. Show that the total power radiated through a sphere of large radius is $2Q_0^2 l^4\omega^6/c^4$.*

We shall first find the radiative magnetic field produced by a dipole formed by $+Q$ and $-Q$ distant l apart, with the mid-point at the origin. The current in a thin wire joining the charges would be

$$i = \frac{dQ}{dt} = \omega Q_0 \cos \omega t.$$

The vector potential at a point P distant r from the origin

$$A_z = \frac{\mu_0}{4\pi}\frac{[i]l}{r}$$

since l is small.

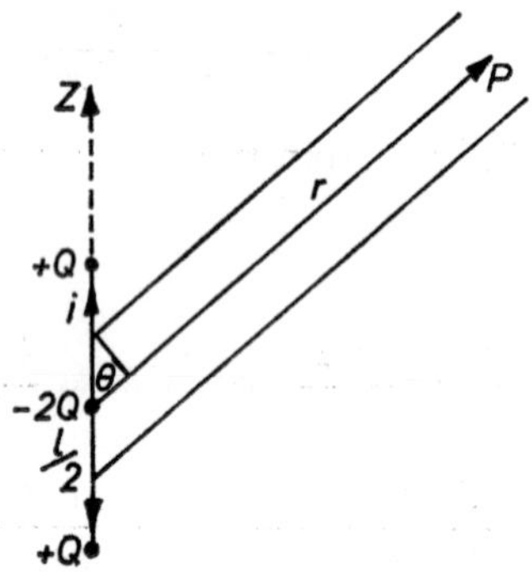

FIG. 45

The square brackets indicate the retarded value of i, that is for a time $t-r/c$.

$$H_x = \frac{1}{\mu_0}\frac{\partial A_z}{\partial y} = \frac{l}{4\pi}\left\{-\frac{y}{r^3}[i]-\frac{y}{cr^2}\left[\frac{\partial i}{\partial t}\right]\right\},$$

$$= -\frac{l}{4\pi}\frac{y}{r}\left\{\frac{[i]}{r^2}+\frac{1}{cr}\left[\frac{\partial i}{\partial t}\right]\right\}.$$

$$H_y = -\frac{1}{\mu_0}\frac{\partial A_z}{\partial x} = \frac{l}{4\pi}\frac{x}{r}\left\{\frac{[i]}{r^2}+\frac{1}{cr}\left[\frac{\partial i}{\partial t}\right]\right\}.$$

$$H = \sqrt{H_x^2+H_y^2} = \frac{l}{4\pi}\sin\theta\left\{\frac{[i]}{r^2}+\frac{1}{cr}\left[\frac{\partial i}{\partial t}\right]\right\}.$$

The radiative magnetic field is given by

$$\frac{l}{4\pi cr}\sin\theta\left[\frac{\partial i}{\partial t}\right] = -\frac{l}{4\pi cr}\sin\theta\,\omega^2 Q_0 \sin\omega\left(t-\frac{r}{c}\right).$$

We can now consider the quadrupole as two dipoles end to end with the centres at $z = \pm l/2$, and currents in opposite directions. The effect of the different distances of the centres from P will be significant only in producing a phase difference at P. These distances will be $r-(l/2)\cos\theta$ and $r+(l/2)\cos\theta$. The resultant field at P will be

$$H = -\left(\frac{\omega^2 Q_0 l \sin\theta}{4\pi c r}\right)\left\{\sin\left[\omega\left(t-\frac{r}{c}\right)+\frac{\omega l\cos\theta}{2c}\right]\right.$$

$$\left. -\sin\left[\omega\left(t-\frac{r}{c}\right)-\frac{\omega l\cos\theta}{2c}\right]\right\},$$

$$= -\left(\frac{\omega^2 Q_0 l \sin\theta}{4\pi c r}\right)\left\{2\cos\omega\left(t-\frac{r}{c}\right)\sin\frac{\omega l\cos\theta}{2c}\right\}.$$

Hence, since l is small,

$$H = -\frac{\omega^3 Q_0 l^2}{4\pi c^2}\frac{\sin\theta\cos\theta}{r}\cos\omega\left(t-\frac{r}{c}\right).$$

The power radiated through a large sphere of radius r

$$= z_0 \int_0^\pi H^2 2\pi r^2 \sin\theta\, d\theta.$$

Since the mean value of

$$\cos^2\omega\left(t-\frac{r}{c}\right)$$

is $\frac{1}{2}$ we have mean power

$$= Z_0\left(\frac{\omega^3 Q_0 l^2}{c^2}\right)^2 \frac{1}{16\pi}\int_0^\pi \sin^3\theta\cos^2\theta\, d\theta.$$

Let $u = \cos\theta$.

The integral

$$= \int_{-1}^{1} u^2(1-u^2)\, du = \tfrac{2}{3}-\tfrac{2}{5} = \tfrac{4}{15}.$$

Mean power

$$= \frac{Z_0}{60\pi}\left(\frac{\omega^3 Q_0 l^2}{c^2}\right)^2.$$

If we substitute $Z_0 = 120\pi\ \Omega$, then power $= 2Q_0^2 l^4\omega^6/c^4$ W.

8.9. *An unpolarised beam of electromagnetic waves of intensity I_0 is incident on a small volume of gas containing n free electrons. Show that the intensity of the scattered beam at a point distant r in a direction making an angle ϕ with the incident beam is given by*

$$I = \tfrac{1}{2} I_0 n \left(\frac{\mu_0 e^2}{4\pi m r}\right)^2 (1+\cos^2 \phi).$$

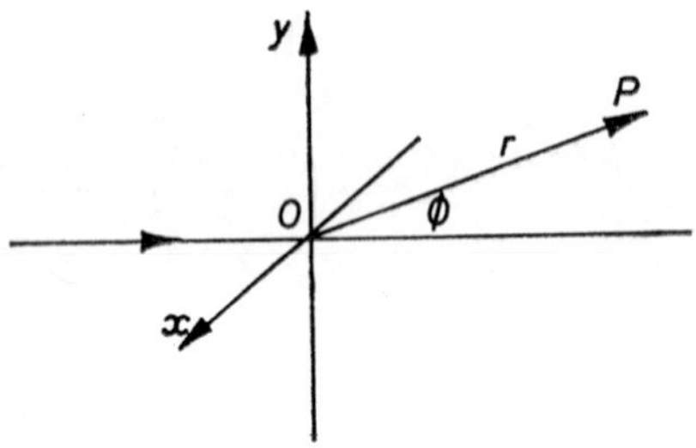

FIG. 46

The radiative magnetic field from an accelerated electron may be taken as $(e \sin \theta / 4\pi c r)\dot{\boldsymbol{v}}$ where θ is the angle between $\dot{\boldsymbol{v}}$ and $\boldsymbol{r}$. $(v \ll c)$.

An unpolarised beam of electromagnetic waves may be represented by a wave in which the electric field $\mathscr{E} \sin \omega t$ can have any direction in a plane normal to the direction of propagation. Since all such directions are equally probable we can represent the field by $(\mathscr{E}/\sqrt{2}) \sin \omega t$ in two directions at right angles. Let one direction (Oy) be chosen in the plane through the beam and the point P, distance r from the small volume containing the electrons, and the other direction (Ox) perpendicular to this. In each direction the acceleration of an electron at any instant will be given by $(e\mathscr{E}/\sqrt{2}m) \sin \omega t$. For the acceleration in the x-direction the field at $\boldsymbol{P}$ is

$$\frac{e^2 \mathscr{E}}{\sqrt{2}m} \frac{\sin \omega t}{4\pi c r},$$

since $\theta = \pi/2$, and for the acceleration in the y-direction the field is

$$-\frac{e^2 \mathscr{E}}{\sqrt{2}m} \frac{\sin \omega t}{4\pi c r} \cos \phi,$$

since

$$\theta = \frac{\pi}{2} - \phi.$$

If H is the resultant field

$$H^2 = \tfrac{1}{2}\left(\frac{e^2}{4\pi crm}\right)^2 \mathscr{E}^2 \sin^2 \omega t(1+\cos^2 \phi).$$

If E = r.m.s. electric field, the mean value of H^2,

$$\overline{H^2} = \tfrac{1}{2}\left(\frac{e^2}{4\pi crm}\right)^2 E^2(1+\cos^2 \phi)$$

Intensity at $P = I = nZ_0\overline{H^2}$ (for n electrons).

Incident intensity $I_0 = E^2/Z_0$.

Therefore

$$I = \tfrac{1}{2}nI_0\left(\frac{e^2}{4\pi crm}\right)^2 Z_0^2(1+\cos^2 \phi).$$

Since

$$Z_0^2 = \frac{\mu_0}{\epsilon_0} \quad \text{and} \quad c^2 = \frac{1}{\mu_0 \epsilon_0}, \quad \frac{Z_0^2}{c^2} = \mu_0^2,$$

$$I = \tfrac{1}{2}nI_0\left(\frac{\mu_0 e^2}{4\pi mr}\right)^2 (1+\cos^2 \phi).$$

EXERCISES

1. A plane wave polarised with the electric field in the plane of incidence is incident upon the surface of a perfect conductor. Discuss the nature of the electric field distribution resulting from the combination of the incident and reflected wave systems.

2. A plane polarised wave of angular frequency ω, travelling *in vacuo* is incident normally on a slab of dielectric of thickness d. Show that the ratio of the amplitude of the electric field in the wave reflected from the slab to that in the incident wave is

$$\left[1+\frac{4v^2c^2 \operatorname{cosec}^2 \theta}{(c^2-v^2)^2}\right]^{-\frac{1}{2}},$$

where v is the velocity of the wave in the dielectric and $\theta = \omega\, d/v$.

3. Show that for a plane electromagnetic wave in a conducting medium the mean energy density is

$$\tfrac{1}{2}\epsilon E_{\text{r.m.s.}}^2\left[1+\left(1+\frac{\sigma^2}{\omega^2\epsilon^2}\right)^{\frac{1}{2}}\right],$$

and the mean value of the Poynting vector is equal to the mean energy density multiplied by the phase velocity.

4. Show that for small conductivity the attenuation constant approaches an upper limit as the frequency is increased.

Compare the attenuation of radiation of wavelength 10 cm in fresh water (conductivity $10^{-3}\ \Omega^{-1}\ \text{m}^{-1}$) and in sea water (conductivity $4\ \Omega^{-1}\ \text{m}^{-1}$). Take the dielectric constant of water to be 81 in both cases.

5. Show that when electromagnetic waves are propagated in a medium of high conductivity the phase and group velocities are given by ω/k and $2\omega/k$ where ω is the angular frequency and k is the attenuation constant.

6. A cylindrical copper conductor of radius 5 mm is employed for the transmission of alternating current of frequency 10 MHz. If the conductivity of copper is $6\times 10^7\ \Omega^{-1}\ \text{m}^{-1}$ find the ratio of its effective resistance at this frequency to its D.C. resistance.

7. The mean rate of absorption of radiant energy from the sun on the earth's surface is $1\cdot 4\ \text{kW m}^{-2}$.

(*a*) Find the r.m.s. value of the electric field in sunlight.

(*b*) Find the total force exerted by solar radiation on the earth.

8. A beamed radio transmission of frequency 50 MHz is found to give no reflection from the ionosphere if the direction of transmission is less than 30° from the vertical. Calculate the maximum electron density in the ionised layer.

9. If W is the mean power radiated by a small current element mounted vertically, show that the r.m.s. electric field at a distance d in a horizontal plane through the element is

$$(3WZ_0/8\pi d^2)^{\frac{1}{2}}$$

where Z_0 is the wave impedance of the medium.

10. Establish expressions for the electric and magnetic fields radiated from a half-wavelength antenna.

Discuss possible arrangements for the production of a narrow beam of radiation in a specified direction. Provide calculations of the angular width of the beam attained.